Power Calculation

by examples

Xing Zhou

Math for Gifted Students

http://www.mathallstar.org

use your mobile device to scan this QR code for more resources including books, practice problems, online courses, and blog.

This book was produced using the LaTeX system.

Contents

CONTENTS

Preface

Welcome to Math All Star© series!

Math All Star originates from a series of lectures given to a group of gifted middle school students with a love for mathematics and an interest in participating in competitions such as MathCounts, AMC, and AIME. These lectures aim to strengthen their problem-solving abilities and to introduce effective techniques that are not typically taught in the classroom.

As the popularity of Math All Star grew, the author began to upload lecture materials to create online courses, thereby providing students with the opportunity to progress at their own paces.

Since then, course materials have constantly been reviewed and updated to reflect student feedback and the observations made during lectures. Recent competition problems are also continuously analyzed and referenced to ensure the relevance of the contents. These course materials are the foundations of this Math All Star series.

Because competition math is a diversified subject that covers both a wide breadth and depth of topics, it is quite challenging to effectively cover all the material in one book that is appropriate for every interested student. Consequently, the author has decided to write a series of books, with each one focusing on a particular topic. Students are encouraged to pick and choose where to begin, depending on their individual skill levels and needs.

In addition to these books, the Math All Star website provides extra practice problems and serves as a highly recommended supplemental learning resource.

If there are any questions, comments, or concerns, please visit the website or email `contact@mathallstar.org`.

Happy learning!

To visit the Math All Star website, scan this QR code or go directly to
`http://www.mathallstar.org`

Introduction

Being able to calculate correctly and efficiently is one of the most basic math skills. Therefore, it is not surprising that various types of calculation problems appear often in math competitions.

This book focuses on basic calculation skills. Contents are grouped into lectures. Each lecture features a classic problem as an example to illustrate one or more techniques. Many of such problems are not too hard to solve. However understanding how to solve these problems is just the very first step. It is more important to understand and master relevant calculation techniques. They can help solve various problems of similar nature in future math competitions.

Readers should have basic algebraic knowledge and skills in order to appreciate the contents in this book. Some problems involve trigonometry and complex number. They can be safely skipped by those readers who are not familiar with these two subjects. Skipping these topics will not affect learning other techniques contained in this book.

Remember: the most important thing is to learn the calculation techniques so you can solve this type of problems, not just one problem at hand!

Lecture 1

$1 + 2 + 3 + \cdots + n$

1.1 Objectives

- The sum formula of an arithmetic sequence

- The sum formula of first n positive integers

- The sum of the first n odd integers

- The reversing order method

1.2 Examples

One of the simplest problems is to find the sum of some *consecutive* integers. For example:

Example 1.2.1

Compute $1 + 2 + \cdots + 2015 + 2016$.

Such a series of numbers is called an *arithmetic sequence*. A sequence is arithmetic if the difference between any two adjacent terms is a constant. This difference is called the *common difference* which is often written as d. In the previous example, d equals 1. However a common difference can be any value as long as it is a constant.

The sum formula for an arithmetic sequence is readily available. This formula can be derived by using the *reversing order method*.

The Reversing Order Method

Let
$$S = 1 + 2 + \cdots + 2015 + 2016 \qquad (1.1)$$

It is obvious that the sum will stay the same if the order of those items on the right of *Equation 1.1* is reversed.

$$
\begin{array}{ccccccccc}
S & = & 1 & + & 2 & + & \cdots & + & 2015 & + & 2016 \\
S & = & 2016 & + & 2015 & + & \cdots & + & 2 & + & 1
\end{array}
$$

Adding the above two equations leads to:

$$2S = (1 + 2016) + (2 + 2015) + \cdots + (2015 + 2) + (2016 + 1)$$
$$= \underbrace{2017 + 2017 + \cdots + 2017}_{2016 \ terms \ in \ total}$$
$$= 2017 \times 2016$$

$$\therefore \quad S = \frac{2017 \times 2016}{2} = 2033136$$

Done.

During a test, the final answer must be given, e.g. 2033136. However while learning, it is acceptable to skip the final calculation if the result is a big number. For example, it is sufficient to give $\frac{2017 \times 2016}{2}$ in this case.

The reversing order method works because the sums of pairing terms are equal. An arithmetic sequence has this property. There are other situations where it is possible to find such pairing terms. The reversing order may be a good fit in those scenarios. Some of such examples will be given in practice and later lectures.

The result of *Example 1.2.1* can be generalized to the following formula.

Formula 1.2.1   Sum of First n Consecutive Integers

$$1 + 2 + 3 + \cdots + n = \frac{(n+1) \cdot n}{2}$$

Formula 1.2.1 can be further generalized to compute the sum of any arithmetic sequence, $a_1, a_2, \cdots, a_n$. This is shown below.

Formula 1.2.2   Sum of Arithmetic Sequence

$$a_1 + a_2 + a_3 + \cdots + a_n = \frac{(a_1 + a_n) \cdot n}{2}$$

This means that the sum of an arithmetic sequence always equals the sum of its 1st and its last term, multiplying by the number of terms and then dividing by 2.

In addition, the following two conclusions are frequently used when tackling sequence related problems.

> ### Theorem 1.2.3   Average of an Arithmetic Sequence
>
> Given an arithmetic sequence $\{a_n\}$, the average of any odd number of consecutive terms must equal its middle term. It follows that the sum of these terms must equal the product of its middle term and the number of terms.

For example, let S_n be the sum of first n terms in an arithmetic sequence and a_n be its nth term. Then it must hold that

$$S_{2n-1} = a_n \cdot (2n - 1)$$

> ### Formula 1.2.4   The nth Term
>
> Let S_n be the sum of first n terms in *any type* of sequence, and a_n be its nth term. It always hold that
>
> $$a_n = S_n - S_{n-1}$$

1.3   Challenges

1) Compute: $1 + 3 + 5 + \cdots + (2n - 1)$.

2) Use the diagram below to explain the sum of first n odd integers is a perfect square.

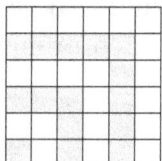

3) Suppose no term in an arithmetic sequence $\{a_n\}$ equals 0. Let S_n be the sum of its first n terms. If $S_{2n-1} = a_n^2$, express its nth term a_n with respect to n.

4) The sum of n different positive integers is less than 100. What is the greatest possible value for n?

5) Determine all pairs (a, b) of real numbers such that $10, a, b, ab$ is an arithmetic progression.

6) A sequence $\{a_n\}$ satisfies $a_n + a_m = a_{n+m}$ for any positive integers n and m. If $a_1 = \frac{1}{2013}$, find the sum of its first 2013 terms.

(Ref 2013 China)

7) Let S_n be the sum of first n terms of an arithmetic sequence. Show that S_n must be in the form of $An^2 + Bn$ where A and B are two constants.

8) Let the sum of first n terms of arithmetic sequence $\{a_n\}$ be S_n, and the sum of first n terms of arithmetic sequence $\{b_n\}$ be T_n. If $S_n : T_n = 2n : 3n + 7$, compute the value of $a_8 : b_6$.

(Ref 2013 China)

9) If the coefficients of the 5th, 6th and 7th terms in the expanded form of $(x^{-\frac{4}{3}} + x)^n$ form an arithmetic sequence, find the constant term in the expanded form.

10) Given a sequence $\{a_n\}$, if $a_n \neq 0$, $a_1 = 1$, and $3a_n a_{n-1} + a_n - a_{n-1} = 0$ for any $n \geq 2$, find the general term of a_n.

Lecture 2

$1 + 2 + 4 + \cdots + 2^n$

2.1 Objectives

- The sum formula of a geometric sequence
- The cancelable term method

2.2 Examples

In *Lecture 1*, we have discussed how to sum an arithmetic sequence. This section will discuss the sum of a *geometric sequence*. A sequence is geometric if the ratio of two adjacent terms is a constant. This ratio is called this sequence's *common ratio*.

Example 2.2.1

Find the sum of $1 + 2 + 4 + \cdots + 2^9 + 2^{10}$.

Finding the sum of a geometric sequence demands a new technique. The idea is to transform the sequence into such a form that the vast majority of intermediate terms can be canceled out.

The Cancelable Term Method

Let

$$S = 1 + 2 + 4 + \cdots + 2^9 + 2^{10} \qquad (2.1)$$

Multiplying both sides of *Equation 2.1* by the common ratio:

$$
\begin{array}{rcccccccccc}
S & = & 1 & + & 2 & + & 4 & + & \cdots & + & 2^9 & + & 2^{10} \\
2 \cdot S & = & & & 2 & + & 4 & + & \cdots & + & 2^9 & + & 2^{10} & + & 2^{11}
\end{array}
$$

For convenience, the right side of the 2nd equation is shifted rightward by one position so that it is clearer to see some terms are equal.

Subtracting the 2nd equation from the 1st one:

$$(1-2) \cdot S = 1 - 2^{11} \implies S = \frac{1 - 2^{11}}{1 - 2} = 2047$$

Done.

The result of *Example 2.2.1* can be generalized into the following formula.

Formula 2.2.1   Sum of Geometric Sequence

The sum of a geometric sequence is given by

$$S = a_1 \cdot \frac{1 - r^n}{1 - r}$$

where a_1 is the first term, r is the common ratio, and n is the number of terms.

In the previous example, $a_1 = 1$, $r = 2$, and there are 11 terms in total.

Caution: Be extra carefully when counting the number of terms in a geometric sequence!

2.3   Challenges

1) Is it possible for a sequence to be both arithmetic and geometric?

2) Simplify the expression: $1 + x + x^2 + \cdots + x^n$.

3) Explain the following polynomial factorization identity using the concept of geometric sequence:

$$1 - x^n = (1 - x)(1 + x + x^2 + \cdots + x^{n-1})$$

4) Compute $2 + 2^3 + 2^5 + \cdots + 2^{2n-1}$.

5) Let $c_1, c_2, c_3, \cdots$ be a series of concentric circles whose radii form a geometric sequence with common ratio as q. Suppose the areas of rings which are formed by two adjacent circles are $S_1, S_2, S_3, \cdots$. Which statement below is correct regarding the sequence $\{S_n\}$?

(A) It is not a geometric sequence

(B) It is a geometric sequence and its common ratio is q

(C) It is a geometric sequence and its common ratio is q^2

(D) It is a geometric sequence and its common ratio is $q^2 - 1$

6) Suppose all the terms in a geometric sequence $\{a_n\}$ are positive. If $|a_2 - a_3| = 14$ and $|a_1 a_2 a_3| = 343$, find a_5.

7) Let sequence $\{a_n\}$ satisfy $a_0 = 1$ and $a_n = \dfrac{\sqrt{1+a_{n-1}^2}-1}{a_{n-1}}$. Prove $a_n > \dfrac{\pi}{2^{n+2}}$.

(Ref 1990 Hungarian)

Lecture 3

$$x + 2x^2 + 3x^3 + \cdots + nx^n$$

3.1 Objectives

- Combine basic techniques to handle more complex sequences

3.2 Examples

Arithmetic sequence and geometric sequence are two basic types of sequences. Math competitions quite often require one to tackle a sequence which is neither arithmetic nor geometric but can be solved by combining several basic techniques.

Example 3.2.1

Simplify $x + 2x^2 + 3x^3 + \cdots + nx^n$.

This sequence exhibits features of both arithmetic and geometric. Its coefficients form an arithmetic sequence and the variable parts form a geometric sequence. Therefore, it is possible to sum this sequence by applying basic techniques appropriately.

Solution

Let $S = x + 2x^2 + 3x^3 + \cdots + nx^n$, then

$$
\begin{aligned}
S &= x + 2x^2 + 3x^3 + \cdots + nx^n \\
x \cdot S &= x^2 + 2x^3 + \cdots + (n-1)x^n + nx^{n+1}
\end{aligned}
$$

Subtracting the 2nd equation from the 1st one results in:

$$(1 - x)S = x + x^2 + x^3 + \cdots + x^n - nx^{n+1}$$

$$= x \cdot \frac{1 - x^n}{1 - x} - nx^{n+1}$$

$$\therefore \quad S = \frac{x(1 - x^n)}{(1 - x)^2} - \frac{nx^{n+1}}{1 - x} \tag{3.1}$$

Done.

ⓘ *Tip: It is always helpful to verify the result by setting n to some small numbers, e.g. $1, 2$, and 3, to check whether both sides equal.*

For instance, setting $n = 1$ in *(3.1)*. The left side $S = x$. The right side $\frac{x(1-x)}{(1-x)^2} - \frac{x^2}{1-x} = x$

Handling such a sequence is one stepping stone to solve 2015 AMC12B problem 25.

3.3 Challenges

1) A sequence satisfies $a_1 = 3, a_2 = 5$, and $a_{n+2} = a_{n+1} - a_n$ for $n \geq 1$ What is the value of a_{2016}?

2) The sum of the first n terms of sequence $\{a_n\}$ is given by the formula $S_n = n^2 + n + 3$ What is the value of a_{10}?

3) Suppose every term in the sequence

$$1, 2, 1, 2, 2, 2, 1, 2, 2, 2, 2, 2, 1, \cdots$$

is either 1 or 2. If there are exactly $(2k-1)$ *twos* between the kth *one* and the $(k+1)^{th}$ *one*, find the sum of its first 2014 terms.
(Ref 2014 China)

4) Simplify $\frac{1}{2} + \frac{2}{2^2} + \frac{3}{2^3} + \cdots + \frac{n}{2^n}$

5) Let sequence $\{a_n\}$ satisfy $a_1 = 2$ and $a_{n+1} = \frac{2(n+2)}{n+1} a_n$ where $n \in \mathbb{Z}^+$. Compute the value of

$$\frac{a_{2014}}{a_1 + a_2 + \cdots + a_{2013}}$$

(Ref 2014 China)

6) Let α and β be the two roots of the equation $x^2 - x - 1 = 0$. If

$$a_n = \frac{\alpha^n - \beta^n}{\alpha - \beta} \quad (n = 1, 2, \cdots)$$

Show that, for any positive integer n, it always hold

$$a_{n+2} = a_{n+1} + a_n$$

7) Let sequence $\{b_n\}$ satisfy $2b_{n+1} = b_n + 3$ and $b_1 = 5$, express b_n with respect to n.

8) If a sequence $\{a_n\}$ satisfies $a_1 = 1$ and

$$a_{n+1} = \frac{1}{16}\left(1 + 4a_n + \sqrt{1 + 24a_n}\right)$$

, express a_n in terms of n.
(Ref 1981 IMO Shortlist)

Lecture 4

$$\frac{1}{2} + \frac{1}{4} + \frac{1}{8} + \cdots$$

4.1 Objectives

- Conditions for an infinite geometric sequence to converge
- Basic concept of limit
- The equation method

4.2 Examples

The number of terms in a sequence can be either limited or unlimited. When there are limited number of terms, their sum always exits. However, when there are infinite number of terms, their sum may or may not exist. Infinite sequence is an important subject in calculus which is beyond the scope of this book. Here, we will only discuss a special type infinite sequence.

 Caution: Strictly speaking, one must first prove the sum of an infinite sequence exists before attempting to calculate this sum.

That being said, problems that appear in middle school and high school level competitions usually satisfy the criteria for their sums to exist. Consequently, it is usually acceptable to compute the sum directly without proving its existence first.

Example 4.2.1

Find the sum of $\frac{1}{2} + \frac{1}{4} + \cdots + \frac{1}{2^n} + \cdots$

Apply the Concept of Limit

Let S_n be the sum of its first n terms.

$$S_n = \frac{1}{2} + \frac{1}{4} + \cdots + \frac{1}{2^n}$$

Then, the desired sum $S = \frac{1}{2} + \frac{1}{4} + \cdots$ is the value of S_n when n approaches infinity. This is written as

$$S = \lim_{n \to \infty} S_n$$

where "lim" stands for *limit*, and ∞ means *infinity*.

By *Formula 2.2.1*,

$$S_n = \frac{1}{2} + \frac{1}{4} + \cdots + \frac{1}{2^n} = \frac{1}{2} \times \frac{1 - \frac{1}{2^n}}{1 - \frac{1}{2}} = 1 - \frac{1}{2^n}$$

When n approaches infinity, $\frac{1}{2^n}$ will approach 0.

$$\therefore \quad S = \lim_{n \to \infty} S_n = \lim_{n \to \infty} \left(1 - \frac{1}{2^n}\right) = 1 - 0 = 1$$

Done.

The result of *Example 4.2.1* can be explained intuitively using the diagram on the right. If we continuously fill $\frac{1}{2}, \frac{1}{4}, \frac{1}{8}$, and so on, of a square, we will eventually fill the whole square.

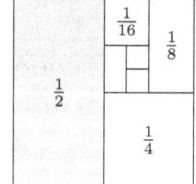

It is apparent that the sequence $\frac{1}{2} + \frac{1}{4} + \frac{1}{8} + \cdots$ has finite sum because the limit of $\frac{1}{2^n}$, or $\left(\frac{1}{2}\right)^n$, exits and diminishes. When the common ratio is 1 or bigger, this will no longer hold. Consequently, the sum of such an infinite sequence will not exist. This conclusion can be summarized as the following.

Theorem 4.2.1   Sum of Infinite Geometric Sequence

The sum of an infinite geometric sequence, $a + ar + ar^2 + ar^3 + \cdots$, exists if and only if its common ratio $|r| < 1$. In such a case, the sum is

$$S = a + ar + ar^2 + ar^3 + \cdots = \frac{a}{1-r}$$

When the sum of an infinite sequence exists, this sequence is said to *converge*. Otherwise, it is sad to *diverge*.

In addition to the above limit based solution, sum of an infinite sequence can also be calculated by using the equation method.

The Equation Method

Let S be the desired sum.

$$S = \frac{1}{2} + \frac{1}{4} + \frac{1}{8} + \cdots$$
$$S - \frac{1}{2} = \frac{1}{4} + \frac{1}{8} + \frac{1}{16} + \cdots$$
$$S - \frac{1}{2} = \frac{1}{2} \times \left(\frac{1}{2} + \frac{1}{4} + \frac{1}{8} + \cdots\right)$$
$$S - \frac{1}{2} = \frac{1}{2} \cdot S$$
$$S = 1$$

Done.

The key to apply this method is to establish an equation by constructing a repetitive pattern. For example, in this example, we note that multiplying $\frac{1}{2}$ with the 2nd term onwards will result in another infinite sequence which is identical to the original one.

The equation method is a useful technique with wider applications beyond summing an infinite geometric sequence. Some of such applications will be discussed in later lectures.

4.3   Challenges

1) Convert $0.\dot{7}$ to a fraction.

2) Convert $0.\dot{7}\dot{3}$ to a fraction.

3) Let

$$f(r) = \sum_{j=2}^{2016} \frac{1}{j^r} = \frac{1}{2^r} + \frac{1}{3^r} + \cdots + \frac{1}{2016^r}$$

   Find

$$\sum_{k=2}^{\infty} f(k)$$

4) Find all x such that $\displaystyle\sum_{k=1}^{\infty} kx^k = 20$.

5) Let $a_1, a_2, \cdots, a_n > 0, n \geq 2$, and $a_1 + a_2 + \cdots + a_n = 1$. Prove

$$\frac{a_1}{2 - a_1} + \frac{a_2}{2 - a_2} + \cdots + \frac{a_n}{2 - a_n} \geq \frac{n}{2n - 1}$$

(Ref 1984 Bulkans) Tip: write each term as the sum of an infinite geometric sequence and then apply the following inequality.

$$\frac{a_1 + a_2 + \cdots + a_n}{n} \leq \sqrt[m]{\frac{a_1^m + a_2^m + \cdots + a_n^m}{n}}$$

Lecture 5

$$\sqrt{2 + \sqrt{2 + \sqrt{2 + \cdots}}}$$

5.1 Objectives

- Simplify infinitely nested radicals
- Review the equation method

5.2 Examples

In addition to infinite geometric sequences, the equation method can tackle other types of infinite expressions too.

Let's consider an infinitely nested radical example now.

Example 5.2.1

Find the value of $\sqrt{2 + \sqrt{2 + \sqrt{2 + \sqrt{2 + \cdots}}}}$.

Such a problem can also be expressed using concepts of sequence and limit which have been discussed in earlier lectures.

If sequence $\{a_n\}$ satisfies

$$a_1 = \sqrt{2}, a_2 = \sqrt{2 + \sqrt{2}}, a_3 = \sqrt{2 + \sqrt{2 + \sqrt{2}}}, ...,$$

find the value of

$$\lim_{n \to \infty} a_n$$

Solution

Let $S = \sqrt{2 + \sqrt{2 + \sqrt{2 + \sqrt{2 + \cdots}}}}$,

then

$$S^2 = 2 + \sqrt{2 + \sqrt{2 + \sqrt{2 + \cdots}}}$$
$$S^2 = 2 + S$$
$$S = 2 \qquad (\because S > 0)$$

Done.

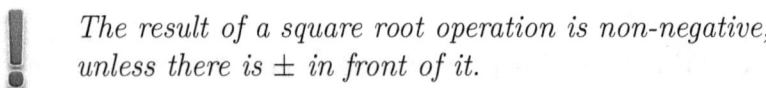 The result of a square root operation is non-negative, unless there is $\pm$ in front of it.

5.3 Challenges

1) Evaluate $\sqrt{7 + \sqrt{2 + \sqrt{2 + \sqrt{2 + \cdots}}}}$.

2) Simplify

$$\cfrac{1}{2 + \cfrac{1}{2 + \cdots}}$$

3) Compute

$$\cfrac{1}{\cfrac{1}{\cdots+1+\frac{1}{\cdots}}+1+\cfrac{1}{\cdots+1+\frac{1}{\cdots}}}$$

4) Simplify

$$(\sqrt{2})^{(\sqrt{2})^{(\sqrt{2})^{\cdots}}}$$

5) Compute

$$\sqrt[3]{6+\sqrt[3]{6+\sqrt[3]{6+\cdots}}}$$

6) Let x be a positive number, use two different reasonings to show that the following relation holds:

$$\sqrt{x\sqrt{x\sqrt{x\sqrt{\cdots}}}}=x$$

7) Find the value of $\sqrt{5+\sqrt{5^2+\sqrt{5^4+\sqrt{5^8+....}}}}$

8) Compute

$$\sqrt{\frac{2}{2^2}+\sqrt{\frac{2}{2^4}+\sqrt{\frac{2}{2^8}+\cdots}}}$$

9) Without using a calculator, explain that

$$\sqrt{20+\sqrt{20+\sqrt{20}}}-\sqrt{20-\sqrt{20-\sqrt{20}}}\approx 1$$

10) Srinivasa Ramanujan, an Indian mathematician, claimed that the following relationship held:

$$x+n=\sqrt{n^2+x\sqrt{n^2+(x+n)\sqrt{n^2+(x+2n)\sqrt{\cdots}}}} \qquad (5.1)$$

Do you agree with him? Explain.

11) Evaluate

$$\sqrt{1+2\sqrt{1+3\sqrt{1+4\sqrt{1+\cdots}}}}$$

Lecture 6

$$\sqrt{2 + \sqrt{3}}$$

6.1 Objectives

- Simplify nested radicals
- Applications of polynomial identities

6.2 Examples

In addition to infinitely nested radicals, some limited nested ones can be simplified too. These problems can usually be solved by applying polynomial square formulas and so on.

Example 6.2.1

Simplify $\sqrt{4 + 2\sqrt{3}}$.

The critical question to ask is, under which circumstance, can such an expression be de-nested? One possibility is that its inner part is a perfect square. If so, the outside radical operator can be removed.

Solution

Suppose that

$$\sqrt{4 + 2\sqrt{3}} = \sqrt{a} + \sqrt{b} \qquad (6.1)$$

Taking square on both sides of the above equation leads to:

$$4 + 2\sqrt{3} = (a + b) + 2\sqrt{ab}$$

Therefore, it must hold that

$$\begin{cases} a + b &= 4 \\ ab &= 3 \end{cases} \implies a, b = 1, 3$$

Hence: $\sqrt{4 + 2\sqrt{3}} = 1 + \sqrt{3}$.

Done.

In order to de-nest an outer square root such as the one in the preceding example, the inside part must be a perfect square. Therefore the key to solve such problem is to find the possible form of the inside part.

The next question is, given $\sqrt{4 + 2\sqrt{3}} = 1 + \sqrt{3}$, how to simplify $\sqrt{2 + \sqrt{3}}$? Apparently, the answer is

$$\sqrt{2 + \sqrt{3}} = \sqrt{\frac{1}{2} \times (4 + 2\sqrt{3})} = \sqrt{\frac{1}{2} \times \left(1 + \sqrt{3}\right)} = \frac{\sqrt{2}}{2} + \frac{\sqrt{6}}{2}$$

To simplify $\sqrt{A + k\sqrt{B}}$, it is usually helpful to first make the coefficient of the inner radical to be 2, i.e. in the form of $\sqrt{A' + 2\sqrt{B'}}$.

Example 6.2.1 is one of the simplest and most straightforward cases. Some of the practice problems will require more careful handling. When relevant, additional notes will be provided in the solutions.

6.3 Challenges

1) Simplify $\sqrt{10 + 4\sqrt{3 - 2\sqrt{2}}}$.

2) Evaluate $\sqrt{7 + 4\sqrt{3}} + \sqrt{7 - 4\sqrt{3}}$.

3) Evaluate
$$\left(\sqrt{3 - 2\sqrt{2}} + \sqrt{3 + 2\sqrt{2}}\right)^2$$

4) Simplify $\sqrt{\sqrt[3]{9} + 6\sqrt[3]{3} + 9}$.

5) Simplify $\sqrt{4 + \sqrt[3]{81} + 4\sqrt[3]{9}}$.

6) Simplify $\sqrt{6 + \sqrt[3]{81} + \sqrt[3]{9}}$.

7) Let $\sqrt{1 + \sqrt{21 + 12\sqrt{3}}} = \sqrt{a} + \sqrt{b}$. Find $a + b$.

8) Solve the equation $x^2 + 6x - 4\sqrt{5} = 0$.

9) Simplify $\sqrt{5\sqrt{3} + 6\sqrt{2}}$.

10) Simplify $(\sqrt[3]{3} + \sqrt[3]{2})(\sqrt[3]{9} - \sqrt[3]{6} + \sqrt[3]{4})$

11) Compute $\sqrt{1 + 1995\sqrt{4 + 1995 \times 1999}}$.

12) Simplify $\sqrt{12 + 2\sqrt{6} + 2\sqrt{14} + 2\sqrt{21}}$.

Lecture 7

$$\frac{1}{1\times2} + \frac{1}{2\times3} + \cdots + \frac{1}{n\times(n+1)}$$

7.1   Objectives

- Review the cancelable term method

7.2   Examples

As shown in the geometric sequence lecture, the cancelable term method is an effective technique to reduce the number of terms. Taking a geometric sequence as an example, regardless of the number of terms in the given sequence, it can always be reduced to just two terms by applying the cancelable term method.

There are several different ways to construct intermediate terms that can be canceled out. Let's consider another example in this lecture.

Example 7.2.1

Compute $\frac{1}{1\times2} + \frac{1}{2\times3} + \cdots + \frac{1}{2016\times2017}$.

27

Solution

Each term can be broken into two by using the identity:

$$\frac{1}{k(k+1)} = \frac{1}{k} - \frac{1}{k+1}$$

$$\therefore \quad \frac{1}{1 \times 2} + \frac{1}{2 \times 3} + \cdots + \frac{1}{2016 \times 2017}$$

$$= \left(1 - \frac{1}{2}\right) + \left(\frac{1}{2} - \frac{1}{3}\right) + \cdots + \left(\frac{1}{2016} - \frac{1}{2017}\right)$$

$$= 1 - \frac{1}{2017}$$

$$= \frac{2016}{2017}$$

Done.

7.3 Challenges

1) Compute the value of the following infinite expression:

$$\frac{1}{1 \times 2} + \frac{1}{2 \times 3} + \cdots$$

2) Suppose function $f(x) = \frac{1+x}{1-x}$. Evaluate

$$f\left(\frac{1}{2}\right) \cdot f\left(\frac{1}{4}\right) \cdot f\left(\frac{1}{6}\right) \cdots f\left(\frac{1}{2014}\right)$$

(Ref 2013 China)

3) Compute

$$\frac{1}{1 \times 3} + \frac{1}{3 \times 5} + \cdots + \frac{1}{2015 \times 2017}$$

4) Compute

$$\sum_{k=1}^{\infty} \frac{1}{k^2 + k}$$

5) Evaluate the infinite sum $\displaystyle\sum_{n=1}^{\infty} \frac{n}{n^4 + 4}$.

6) Let S_n be the sum of first n terms in sequence $\{a_n\}$ where

$$a_n = \sqrt{1 + \frac{1}{n^2} + \frac{1}{(n+1)^2}}$$

Find $\lfloor S_n \rfloor$ where the floor function $\lfloor x \rfloor$ returns the largest integer not exceeding x.

7) Let $\{a_n\}$ be an increasing geometric sequence satisfying $a_1 + a_2 = 6$ and $a_3 + a_4 = 24$. Let $\{b_n\}$ be another sequence satisfying $b_n = \frac{a_n}{(a_n - 1)^2}$. If T_n is the sum of first n terms in $\{b_n\}$, show that for any positive integer n, the relation $T_n < 3$ always holds.

8) For $n \geq 1$, let d_n denote the length of the line segment connecting the two points where the line $y = x + n + 1$ intersects the parabola $8x^2 = y - \frac{1}{32}$. Compute the sum

$$\sum_{n=1}^{1000} \frac{1}{n \cdot d_n^2}$$

9) Compute the value of

$$\sum_{n=1}^{\infty} \frac{2n + 1}{n^2 (n+1)^2}$$

10) Find the length of the leading non-repeating block in the decimal expansion of $\frac{2017}{3 \times 5^{2016}}$. For example the length of the leading non-repeating block of $\frac{1}{6} = 0.1\overline{6}$ is 1.

Lecture 8

$$\frac{1}{1\times2\times3} + \frac{1}{2\times3\times4} + \cdots + \frac{1}{n\times(n+1)\times(n+2)}$$

8.1 Objectives

- More on cancelable terms
- The special value method

8.2 Examples

When there are only two terms, we can show that

$$\frac{1}{k\times(k+1)} = \frac{1}{k} - \frac{1}{k+1}$$

or, more generally (suppose d is a constant)

$$\frac{1}{k\times(k+d)} = \frac{1}{d}\times\left(\frac{1}{k} - \frac{1}{k+d}\right) \qquad (8.1)$$

What if there are three or more terms in the denominator?

Example 8.2.1

Compute $\frac{1}{1\times2\times3} + \frac{1}{2\times3\times4} + \cdots + \frac{1}{8\times9\times10}$.

Solution

If there are three terms in the denominator, it is possible to decompose this fraction into two whose denominators contain two terms each.

$$\frac{1}{1\times2\times3} = \frac{1}{2} \times \left(\frac{1}{1\times2} - \frac{1}{2\times3}\right)$$

$$\frac{1}{2\times3\times4} = \frac{1}{2} \times \left(\frac{1}{2\times3} - \frac{1}{3\times4}\right)$$

$$\cdots$$

$$\frac{1}{8\times9\times10} = \frac{1}{2} \times \left(\frac{1}{8\times9} - \frac{1}{9\times10}\right)$$

Adding these equations together and canceling equal terms yield:

$$\frac{1}{1\times2\times3} + \frac{1}{2\times3\times4} + \cdots + \frac{1}{8\times9\times10}$$
$$= \frac{1}{2} \times \left(\frac{1}{1\times2} - \frac{1}{9\times10}\right)$$
$$= \frac{11}{45}$$

Done.

This is an elegant solution. A similar approach can be used to sum a sequence with more terms in the denominator such as

$$\frac{1}{1\times2\times3\times4} + \frac{1}{2\times3\times4\times5} + \cdots$$

In order to master this technique, it is important to understand how the identities such as the following are constructed.

$$\frac{1}{1 \times 2 \times 3} = \frac{1}{2} \times \left(\frac{1}{1 \times 2} - \frac{1}{2 \times 3}\right) \qquad (8.2)$$

It turns out that *(8.2)* is just an extension of *(8.1)*. All it takes is just to extract the middle term temporarily and apply back afterwards.

$$\frac{1}{1 \times 2 \times 3} = \frac{1}{2}\left(\frac{1}{1 \times 3}\right) = \frac{1}{2}\left(\frac{1}{2}\left(\frac{1}{1} - \frac{1}{3}\right)\right) = \frac{1}{2}\left(\frac{1}{1 \times 2} - \frac{1}{2 \times 3}\right)$$

Similarly,

$$\frac{1}{2 \times 3 \times 4} = \frac{1}{3}\left(\frac{1}{2 \times 4}\right) = \frac{1}{3}\left(\frac{1}{2}\left(\frac{1}{2} - \frac{1}{4}\right)\right) = \frac{1}{2}\left(\frac{1}{2 \times 3} - \frac{1}{3 \times 4}\right)$$

More generally,

$$\frac{1}{n(n+1)(n+2)}$$
$$= \frac{1}{n+1}\left(\frac{1}{n(n+2)}\right)$$
$$= \frac{1}{n+1}\left(\frac{1}{2}\left(\frac{1}{n} - \frac{1}{n+2}\right)\right)$$
$$= \frac{1}{2}\left(\frac{1}{n(n+1)} - \frac{1}{(n+1)(n+2)}\right)$$

> Many competition problems can be solved by using a combination or extensions of basic techniques.

Additionally, *Example 8.2.1* can also be solved by an alternative solution. In this particular case, this approach may appear less elegant than the previous solution does. However, the technique involved is useful and powerful. It can be used to solve many other types of problems.

The Special Value Method

Suppose that

$$\frac{1}{k \times (k+1) \times (k+2)} = \frac{A}{k} + \frac{B}{k+1} + \frac{C}{k+2} \qquad (8.3)$$

where A, B, and C are to be determined constants.

Multiplying both sides of *Equation 8.3* by $k(k+1)(k+2)$ yields:

$$1 = A \cdot (k+1) \cdot (k+2) + B \cdot k \cdot (k+2) + C \cdot k \cdot (k+1) \qquad (8.4)$$

It is possible to determine A, B, and C by expanding the above equation and then comparing corresponding coefficients. However there is a more general and usually easier way.

Because *Equation 8.4* is an identity, the equality will hold regardless of the value of k. Hence we can assign some special values to k in order to solve A, B, and C.

$$\text{Let } k = 0 \quad \Longrightarrow \quad 1 = 2A \quad \Longrightarrow \quad A = \tfrac{1}{2}$$

$$\text{Let } k = -1 \quad \Longrightarrow \quad 1 = -B \quad \Longrightarrow \quad B = -1$$

$$\text{Let } k = -2 \quad \Longrightarrow \quad 1 = 2C \quad \Longrightarrow \quad C = \tfrac{1}{2}$$

Therefore, we conclude

$$\frac{1}{k \times (k+1) \times (k+2)} = \frac{1}{2} \times \frac{1}{k} - 1 \times \frac{1}{k+1} + \frac{1}{2} \times \frac{1}{k+2}$$

This is followed by:

$$\frac{1}{1\times2\times3} = \frac{1}{2}\times\frac{1}{1}+(-1)\times\frac{1}{2}+\boxed{\frac{1}{2}\times\frac{1}{3}}$$

$$\frac{1}{2\times3\times4} = \frac{1}{2}\times\frac{1}{2}+\boxed{(-1)\times\frac{1}{3}}+\frac{1}{2}\times\frac{1}{4}$$

$$\frac{1}{3\times4\times5} = \boxed{\frac{1}{2}\times\frac{1}{3}}+(-1)\times\frac{1}{4}+\frac{1}{2}\times\frac{1}{5}$$

$$\frac{1}{4\times5\times6} = \frac{1}{2}\times\frac{1}{4}+(-1)\times\frac{1}{5}+\frac{1}{2}\times\frac{1}{6}$$

$$\frac{1}{5\times6\times7} = \frac{1}{2}\times\frac{1}{5}+(-1)\times\frac{1}{6}+\frac{1}{2}\times\frac{1}{7}$$

$$\frac{1}{6\times7\times8} = \frac{1}{2}\times\frac{1}{6}+(-1)\times\frac{1}{7}+\boxed{\frac{1}{2}\times\frac{1}{8}}$$

$$\frac{1}{7\times8\times9} = \frac{1}{2}\times\frac{1}{7}+\boxed{(-1)\times\frac{1}{8}}+\frac{1}{2}\times\frac{1}{9}$$

$$\frac{1}{8\times9\times10} = \boxed{\frac{1}{2}\times\frac{1}{8}}+(-1)\times\frac{1}{9}+\frac{1}{2}\times\frac{1}{10}$$

Many intermediate terms will be canceled when these equations are added together. Only the top left triangular corner and the bottom right one will remain. Therefore,

$$
\begin{aligned}
S &= \frac{1}{2}\times\frac{1}{1}-1\times\frac{1}{2}+\frac{1}{2}\times\frac{1}{2}+\frac{1}{2}\times\frac{1}{9}-1\times\frac{1}{9}+\frac{1}{2}\times\frac{1}{10}\\
&= \frac{1}{2}\times\frac{1}{1}-\frac{1}{2}\times\frac{1}{2}-\frac{1}{2}\times\frac{1}{9}+\frac{1}{2}\times\frac{1}{10}\\
&= \frac{1}{4}-\frac{1}{2\times9\times10}\\
&= \frac{11}{45}
\end{aligned}
$$

Done.

The essence of the special value method is to utilize the fact that an identity always holds regardless of the values of variables. Therefore it is possible to derive some useful results by setting variables to some special values. This method will be discussed again in later lectures.

8.3 Challenges

1) Simplify

$$\frac{1}{1 \times 2 \times 3} + \frac{1}{2 \times 3 \times 4} + \cdots + \frac{1}{n \times (n+1) \times (n+2)}$$

2) Compute the value of the following infinite expression:

$$\frac{1}{1 \times 2 \times 3} + \frac{1}{2 \times 3 \times 4} + \cdots$$

3) Use the *special value method* to determine coefficient A and B in the identities below:

$$\frac{1}{k \times (k+1)} = \frac{A}{k} + \frac{B}{k+1}$$

$$\frac{1}{k \times (k+2)} = \frac{A}{k} + \frac{B}{k+2}$$

4) Evaluate the value of

$$\sum_{n=1}^{\infty} \frac{6}{(2n-1)(2n+1)}$$

5) Compute

$$\sum_{n=1}^{\infty} \frac{2}{n^2 + 4n + 3}$$

6) Find the remainder when $x^{81} + x^{49} + x^{25} + x^9 + x$ is divided by $x^3 - x$.

Lecture 9

$$C_n^0 + C_n^1 + \cdots + C_n^n$$

9.1 Objectives

- Pascal triangle's properties
- Binomial theorem
- Review the special value method

9.2 Examples

Combination numbers are closely related to Pascal's triangle and the binomial theorem. Consequently, combinatorics identities can often be understood and solved using these two methods.

Example 9.2.1

Compute the value of $C_n^0 + C_n^1 + \cdots + C_n^n$.

There are several different ways to solve this problem. Two of them will be discussed in this lecture. One uses properties of

Pascal's triangle. The other is based on the binomial theorem.

Pascal's triangle

It is well known that every element in Pascal's triangle is a combination number. If both the row and the column indices start at 0, then the element of the nth row and the kth column equals C_n^k.

Pascal's triangle has many interesting properties. One of them states that the sum of all the elements in n^{th} row is equal to 2^n.

$$
\begin{array}{ccccccccccccc}
& & & & & & 1 & & & & & & & \cdots\cdots & 2^0 \\
& & & & & 1 & & 1 & & & & & & \cdots\cdots & 2^1 \\
& & & & 1 & & 2 & & 1 & & & & & \cdots\cdots & 2^2 \\
& & & 1 & & 3 & & 3 & & 1 & & & & \cdots\cdots & 2^3 \\
& & 1 & & 4 & & 6 & & 4 & & 1 & & & \cdots\cdots & 2^4 \\
& 1 & & 5 & & 10 & & 10 & & 5 & & 1 & & \cdots\cdots & 2^5 \\
1 & & 6 & & 15 & & 20 & & 15 & & 6 & & 1 & \cdots\cdots & 2^6
\end{array}
$$

Theorem 9.2.1 Pascal's Triangle - Row Sum

The sum of all elements of the nth row in Pascal's triangle equals 2^n.

Because elements of the nth row are $C_n^0, C_n^1, \cdots, C_n^n$, therefore their sum must equal 2^n, i.e.

$$C_n^0 + C_n^1 + \cdots + C_n^n = 2^n \tag{9.1}$$

Done.

In addition to relying on Pascal's triangle's properties, this identity can also be proved using the binomial theorem.

> ### Theorem 9.2.2   Binomial Theorem
>
> $$(a+b)^n = C_n^0 a^n + C_n^1 a^{n-1}b + C_n^2 a^{n-2}b^2 + \cdots + C_n^{n-1}ab^{n-1} + C_n^n b^n$$

Proof of this theorem can be found in many textbooks. Thus it will not be discussed here for concise reason.[1]

Because *(9.2)* is an identity, it will always hold regardless of the value of a and b. It is then possible to derive many interesting results by assigning some special values to a and b.

> Assign special values to an identity is a useful technique.

Binomial Based Solution

Setting $a = b = 1$ in *Equation 9.2.2* yields:

$$(1+1)^n = C_n^0 + C_n^1 + C_n^2 + \cdots + C_n^n$$

$$\therefore \quad 2^n = C_n^0 + C_n^1 + C_n^2 + \cdots + C_n^n$$

Done.

> The sum of all coefficients in the expanded form of a polynomial can be computed by assigning value 1 to all variables.

Quite often, the binomial theorem is written as the following form:

$$(1+x)^n = C_n^0 + C_n^1 x + C_n^2 x^2 + \cdots + C_n^n x^n \qquad (9.2)$$

Equation 9.2 can be viewed as the special case of *(9.2.2)* where one of the variables is already assigned a value 1.

---

[1] An intuitive explanation of binomial theorem and its extension, multinomial theorem, is offered in the book *Counting* by the same author.

9.3 Challenges

1) Show that

$$C_n^0 + 2C_n^1 + 4C_n^2 + \cdots + 2^n C_n^n = 3^n$$

2) Compute

$$C_n^1 + 2C_n^2 + \cdots + nC_n^n$$

3) If the sum of all coefficients in the expanded form of $(3x + 1)^n$ is 256, find the coefficient of x^2.

4) Find the coefficient of x^{17} in the expansion of $(1 + x^5 + x^7)^{20}$.

   (Ref 2001 HK Team Selection)

5) **(Vandermonde Identity)** Given positive integers m, n, and r, where $r \leq m$ and $r \leq n$, prove the following relation always holds:

$$C_m^0 C_n^r + C_m^1 C_n^{r-1} + \cdots + C_m^r C_n^0 = C_{m+n}^r$$

6) Simplify $\sum_{k=0}^{n} \left(C_n^k\right)^2 = \left(C_n^0\right)^2 + \left(C_n^1\right)^2 + \cdots + \left(C_n^n\right)^2$.

7) Show that

$$\sum_{i=1}^{n} C_n^i C_n^{i-1} = C_{2n}^{n-1}$$

8) Use the binomial theorem to prove that $3^{4n+2} + 5^{2n+1}$ is divisible by 14 for any given positive integer n.

9) Let the integer and decimal part of $(5\sqrt{2} + 7)^{2n+1}$ be I and D respectively. Show that $(I + D) \cdot D$ is a constant.

10) Let a, b be two positive real numbers, and n be a positive integer greater than 2. Show that

$$\frac{a^n + a^{n-1}b + \cdots + ab^{-1} + b^n}{n + 1} \geq \left(\frac{a + b}{2}\right)^n$$

Lecture 10

$$C_n^0 + C_n^2 + C_n^4 + \cdots$$

10.1 Objectives

- More on the special value method
- Review the binomial theorem

10.2 Examples

In the previous lecture, we have discussed how to compute the sum of all coefficients of an expanded polynomial. In this lecture we will expand this topic and derive some more interesting results.

Example 10.2.1

Compute
$$C_{2016}^0 + C_{2016}^2 + \cdots + C_{2016}^{2016}$$
and
$$C_{2016}^1 + C_{2016}^3 + \cdots + C_{2016}^{2015}$$

Such problems can also be solved by assigning different values to x in $(1+x)^n$ and its expanded form.

Solution

In identity $(1+x)^n = C_n^0 + C_n^1 x + C_n^2 x + \cdots + C_n^{n-1} x^{n-1} + C_n^n x^n$,

let $x = 1 \implies (1+1)^n = C_n^0 + C_n^1 + C_n^2 + \cdots$

let $x = -1 \implies (1-1)^n = C_n^0 - C_n^1 + C_n^2 - \cdots$

In the 1st equation, all coefficients are 1. In the 2nd one, all the odd terms' coefficients are 1 and all the even terms' are -1.

Adding these two equations together will cancel all the even terms.

$$(1+1)^n + (1-1)^n = 2(C_n^0 + C_n^2 + \cdots) \implies \boxed{C_n^0 + C_n^2 + \cdots = 2^{n-1}}$$

Subtracting the 2nd equation from the the 1st one will cancel all the odd terms.

$$(1+1)^n - (1-1)^n = 2(C_n^1 + C_n^3 + \cdots) \implies \boxed{C_n^1 + C_n^3 + \cdots = 2^{n-1}}$$

By setting $n = 2016$, we find the answers to both of the original questions are $\boxed{2^{2015}}$.

Done.

The key to apply the special value method is to decide which special values to use.

Therefore, in order to be proficient in solving such type of questions, one must understand why ± 1 are chosen in this case.

A common feature in both *Example 9.2.1* and *Example 10.2.1* is their "superscripts" form an arithmetic sequence:

Example 9.2.1:

$$C_n^0 + C_n^1 + C_n^2 + \cdots \qquad (10.1)$$

Example 10.2.1:

$$C_n^0 + C_n^2 + C_n^4 + \cdots \qquad (10.2)$$
$$C_n^1 + C_n^3 + C_n^5 + \cdots \qquad (10.3)$$

Similarly, one may also be asked to compute:

$$C_n^0 + C_n^3 + C_n^6 + \cdots = \quad ? \qquad (10.4)$$
$$C_n^0 + C_n^4 + C_n^8 + \cdots = \quad ? \qquad (10.5)$$

$$\cdots$$

> Such problems can all be solved by assigning the complex roots of $x^d = 1$ to x where d is the common difference of the arithmetic sequence formed by their superscripts.

For instance,

- To solve *(10.1)*, set $x = 1$ which is the only root of $x^1 = 1$

- To solve *(10.2)*, set $x = \pm 1$ which are the two roots of $x^2 = 1$

Similarly, *(10.4)* can be solved by setting $x = 1, \omega$, and ω^2, respectively, where $\omega = \cos \frac{2\pi}{3} + i \sin \frac{2\pi}{3}$. They are the three roots of $x^3 = 1$.

To solve *(10.5)*, we can set x to $1, i, -1$, and $-i$, respectively. They are the four roots of $x^4 = 1$.

10.3 Challenges

1) Evaluate

$$C_n^0 - C_n^1 + C_n^2 - C_n^3 + \cdots \qquad (10.6)$$

2) Show that

$$C_n^0 - C_n^2 + C_n^4 - C_n^6 + \cdots = 2^{\frac{n}{2}} \cos \frac{n\pi}{4}$$
$$C_n^1 - C_n^3 + C_n^5 - C_n^7 + \cdots = 2^{\frac{n}{2}} \sin \frac{n\pi}{4}$$

3) Simplify the following expressions:

(i) $C_{2016}^0 + C_{2016}^3 + C_{2016}^6 + \cdots + C_{2016}^{2016}$

(ii) $C_{2016}^1 + C_{2016}^4 + C_{2016}^7 + \cdots + C_{2016}^{2014}$

(iii) $C_{2016}^2 + C_{2016}^5 + C_{2016}^8 + \cdots + C_{2016}^{2016}$

4) If $a_n = C_{2003}^{3n-1}$, find the value of $\displaystyle\sum_{n=1}^{668} a_n$.

(Ref 2003 China)

5) Simplify

$$C_n^0 - \frac{1}{2}C_n^1 + C_n^2 - \frac{1}{2}C_n^3 + \cdots$$

Lecture 11

$$C_k^k + C_{k+1}^k + C_{k+2}^k + \cdots + C_n^k$$

11.1 Objectives

- The Hockey Stick identity
- More on Pascal's triangle's property

11.2 Examples

The hockey stick identity is well known among strong math competition contenders. It states that

---

Theorem 11.2.1 Hockey Stick Identity

$$C_k^k + C_{k+1}^k + C_{k+2}^k \cdots + C_n^k = C_{n+1}^{k+1}$$

The first term on the left always has equal subscript and superscript. Subsequent terms always have the same superscript but increasing subscripts. The result's subscript and superscript equal the values of those of the last term on the left plus one.

This identity can also be remembered and intuitively explained using Pascal's triangle. The diagram below describes the following relationship:

$$C_2^2 + C_3^2 + C_4^2 + C_5^2 = C_6^3$$

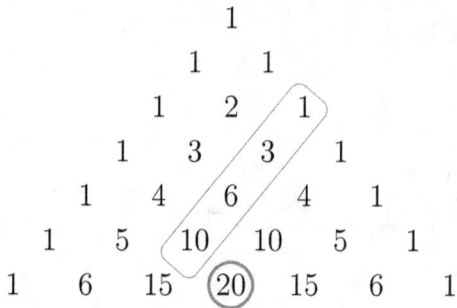

The sequence can start from any point on the edge of Pascal's triangle. It can go parallel to the other edge for as long as it wants. The sum of this sequence can be found by turning 90° downward from the last node of this sequence.

The validity of the hockey stick identity can be verified using Pascal's triangle in an intuitive way. We note that all numbers on edges of Pascal's triangle equal 1. Therefore, it is safe to just move the starting node downward by one position along the edge, as shown below.

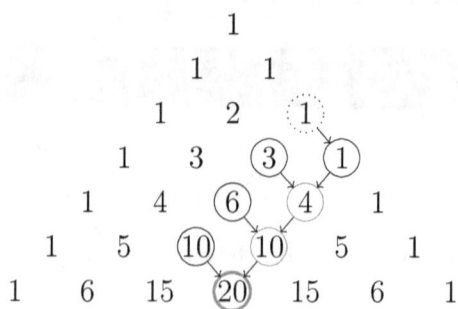

Then, by construction of Pascal's triangle, a repetitive process can be started by replaced the two shoulder nodes with the one below. This process can go as long as it needs to be. The diagram above represents the following relation:

$$1 + 3 + 6 + 10 = ((1 + 3) + 6) + 10 = (4 + 6) + 10 = 10 + 10 = 20$$

or

$$C_2^2 + C_3^2 + C_4^2 + C_5^2 = C_6^3$$

Hockey stick is one of most tested combinatorial identities. For example, the $12th$ problem in 2015 AIME I can be solved using this identity.

11.3 Challenges

1) When preparing goodie bags for his birthday party, Joe randomly puts between 20 and 30 M&M chocolates in each bag. Suppose M&M has 5 different colors available, how many different goodie bags are possible?

2) Let $n > k$ be two positive integers. Simplify

$$C_n^k + 2C_{n-1}^k + 3C_{n-2}^k + \cdots (n - k + 1)C_k^k$$

3) Using the Hockey Stick identity to compute $1 + 2 + 3 + \cdots + n$.

4) Evaluate $1 \times 2 + 2 \times 3 + 3 \times 4 + \cdots + n(n + 1)$.

5) Compute $1 \times 2 \times 3 + 2 \times 3 \times 4 + \cdots + 2016 \times 2017 \times 2018$.

Lecture 12

$$1 \times 2 + 2 \times 3 + \cdots + n \times (n+1)$$

12.1 Objectives

- Application of hockey stick identity
- The mathematical induction
- The sum of square formula

12.2 Examples

Let's revisit the sequence

$$1 + 2 + 3 + \cdots + n \tag{12.1}$$

This is an arithmetic sequence which was discussed in *Lecture 1*. Meanwhile *(12.1)* can also be viewed as the simplest case of other families of more complex sequences. For example, the following series are to sum products of several consecutive integers.

$$S_1 = 1 + 2 + 3 + \cdots + n$$
$$S_2 = 1 \times 2 + 2 \times 3 + 3 \times 4 + \cdots + n(n+1)$$
$$S_3 = 1 \times 2 \times 3 + 2 \times 3 \times 4 + 3 \times 4 \times 5 + \cdots + n(n+1)(n+2)$$
$$\cdots$$

Apparently, *(12.1)* is the simplest one in this family. However, except S_1, none of the others is an arithmetic sequences and cannot be handled by the reversing order method

Meanwhile, *(12.1)* is also the simplest one in the following power series family:

$$P_1 = 1 + 2 + 3 + \cdots + n$$
$$P_2 = 1^2 + 2^2 + 3^2 + \cdots + n^2$$
$$P_3 = 1^3 + 2^3 + 3^3 + \cdots + n^3$$
$$\cdots$$

Again, except P_1, none of them is arithmetic and thus cannot be computed by the reversing order method either.

This lecture will discuss how to compute these two families of sequences. Please note that *Example 12.2.1* is one of the practice problems in the previous lecture. A full solution will be given here.

Example 12.2.1

Compute $1 \times 2 + 2 \times 3 + 3 \times 4 + \cdots + n(n+1)$.

It is possible to find the result by expanding each term. Using this approach, the original expression can be re-written as

$$1 \times 2 + 2 \times 3 + 3 \times 4 + \cdots + n(n+1)$$
$$= 1 \times (1+1) + 2 \times (2+1) + 3 \times (3+1) + \cdots + n(n+1)$$
$$= (1^2 + 1) + (2^2 + 2) + (3^3 + 3) + \cdots + (n^2 + n)$$
$$= (1 + 2 + \cdots + n) + (1^2 + 2^2 + \cdots + n^2)$$

This approach works if both the sum of an arithmetic sequence and the sum of squares are known. When the number of consecutive integers in each term increases, such approach will require the sum formula of higher power. For example, if each term contains 4 consecutive integers, sum of 4th powers will be required.

$$1 \times 2 \times 3 \times 4 + \cdots = (1 + 2 + \cdots) + \cdots + (1^4 + 2^4 + \cdots)$$

Apparently, this approach is not scalable. A better alternative is to apply the hockey stick identity which has been discussed in *Lecture 11*.

Solution

$$1 \times 2 + 2 \times 3 + 3 \times 4 + \cdots + n(n+1)$$
$$=2! \times \left(\frac{1 \times 2}{2!} + \frac{2 \times 3}{2!} + \frac{3 \times 4}{2!} + \cdots + \frac{n(n+1)}{2!} \right)$$
$$=2 \times (C_2^2 + C_3^2 + C_4^2 + \cdots + C_{n+1}^2)$$
$$=2 \times C_{n+2}^3 \qquad \qquad \because \text{ Hockey stick Idetity}$$
$$=\frac{n(n+1)(n+2)}{3}$$

Done.

Clearly, this solution is scalable because the number of consecutive integers in each term no longer presents an obstacle.

Now, let's consider the sum of a power series.

Example 12.2.2

Derive the formula of $1^2 + 2^2 + \cdots + n^2$.

The sum of squares formula is well known among some students.

$$1^2 + 2^2 + \cdots + n^2 = \frac{1}{6}n(n+1)(2n+1) \qquad (12.2)$$

This formula is usually proved by using the mathematical induction. In fact, proving the sum of square formula is a classic example of mathematical induction which can be found in many textbooks.

Mathematical Induction Based Solution

When $n = 1$, the left side of *Formula 12.2* equals 1 and the right side equals 1 too. Therefore, the relation holds when $n = 1$.

Assuming it holds when $n = k$, let's show that it will hold for $n = k + 1$ as well.

$$1^2 + 2^2 + \cdots + k^2 + (k+1)^2$$
$$= \frac{1}{6}k(k+1)(2k+1) + (k+1)^2 \qquad \because \text{ the assumption of } n = k$$
$$= \frac{1}{6}(k+1)\Big(k(2k+1) + 6(k+1)\Big)$$
$$= \frac{1}{6}(k+1)(2k^2 + 7k + 6)$$
$$= \frac{1}{6}(k+1)(2k+3)(k+2)$$
$$= \frac{1}{6}(k+1)((k+1)+1)(2(k+1)+1)$$

Therefore by the principle of mathematical induction, *(12.2)* holds for all positive integer n.

Done.

Those readers who are not familiar with the principle of mathematical induction can read a suitable textbook. Many textbooks contain in depth description.

While it is a powerful method, mathematical induction requires the result to be known in advance. In another word, it can prove the result is correct, but usually cannot derive the result directly. Hence, a question that rises naturally is how to derive the result without knowing it beforehand?

Finding the result can be done by guessing or pattern seeking. Often, it can also be achieved by employing appropriate calculation techniques. For example, the sum of a power series can be computed directly. This is shown below.

Solution

Because $k^2 = k(k+1) - k$, we have

$$1^2 + 2^2 + \cdots + n^2$$
$$=(1 \times 2 - 1) + (2 \times 3 - 2) + \cdots + (n \times (n+1) - n)$$
$$=(1 \times 2 + 2 \times 3 + \cdots + n \times (n+1)) - (1 + 2 + \cdots + n)$$
$$=\frac{1}{3}n(n+1)(n+2) - \frac{1}{2}n(n+1) \qquad \because \text{Example 12.2.1}$$
$$=\frac{1}{6}n(n+1)(2n+1)$$

Done.

12.3 Challenges

1) Simplify $1 + 3 + 6 + \cdots + \frac{n(n+1)}{2}$.

2) Explain the principle of mathematical induction.

3) Using the principle of mathematical induction to prove

$$1^3 + 2^3 + \cdots + n^3 = \left[\frac{n(n+1)}{2}\right]^2 = (1 + 2 + \cdots + n)^2$$

4) Show that $1(1!) + 2(2!) + 3(3!) + \cdots + n(n!) = (n+1)! - 1$

5) Simplify $3^3 + 6^3 + 9^3 + \cdots + (3n)^3$.

6) Find the sum formula $1^3 + 2^3 + \cdots + n^3$ directly.

7) Simplify $1^2 + 3^2 + \cdots + (2n-1)^2$.

8) Let n be a positive integer. Show that

$$\left(1+\frac{1}{3}\right)\left(1+\frac{1}{3^2}\right)\cdots\left(1+\frac{1}{3^n}\right) < 2$$

(Ref 2012 China)

Lecture 13

$$C_n^0 + \frac{1}{2}C_n^1 + \frac{1}{3}C_n^2 + \cdots + \frac{1}{n+1}C_n^n$$

13.1 Objectives

- Basic combinatorial identities

13.2 Examples

Combinatorial identities appear frequently in various competitions. There are hundreds of, if not more, combinatorial identities. It is impossible and unnecessary to remember all of them. That being said, it is important to remember those most frequently used and basic ones.

Theorem 13.2.1 Basic Combinatorial Identities

$$C_n^k + C_n^{k+1} = C_{n+1}^{k+1} \tag{13.1}$$

$$C_n^k = \frac{k+1}{n+1}C_{n+1}^{k+1} \tag{13.2}$$

Proof

Both identities can be proved by applying combination number's definition directly.

$$(13.1) \quad C_n^k + C_n^{k+1} = \frac{n!}{k!(n-k)!} + \frac{n!}{(k+1)!(n-k-1)!}$$
$$= \frac{n!(k+1) + n!(n-k)}{(k+1)!(n-k)!}$$
$$= \frac{(n+1)!}{(k+1)!(n-k)!}$$
$$= C_{n+1}^{k+1}$$

$$(13.2) \quad \frac{k+1}{n+1} C_{n+1}^{k+1} = \frac{k+1}{n+1} \frac{(n+1)!}{(k+1)!(n-k)!}$$
$$= \frac{n!}{k!(n-k)!}$$
$$= C_n^k$$

QED

Both *(13.1)* and *(13.1)* are fundamental stepping stones to derive many interesting combinatorial relations. The following example is one of them.

Example 13.2.1

Simplify

$$C_n^0 + \frac{1}{2}C_n^1 + \frac{1}{3}C_n^2 + \frac{1}{4}C_n^3 + \cdots + \frac{1}{n+1}C_n^n$$

Solution

Formula 13.2 can be rewritten as

$$\frac{1}{k+1}C_n^k = \frac{1}{n+1}C_{n+1}^{k+1}$$

$$\therefore \quad C_n^0 + \frac{1}{2}C_n^1 + \frac{1}{3}C_n^2 + \frac{1}{4}C_n^3 + \cdots + \frac{1}{n+1}C_n^n$$

$$= \frac{1}{n+1}C_{n+1}^1 + \frac{1}{n+1}C_{n+1}^2 + \frac{1}{n+1}C_{n+1}^3 + \cdots + \frac{1}{n+1}C_{n+1}^{n+1}$$

$$= \frac{1}{n+1}(C_{n+1}^1 + C_{n+1}^2 + \cdots + C_{n+1}^{n+1})$$

$$= \frac{1}{n+1}(2^{n+1} - 1) \qquad \text{by \textit{Equation 9.1} on page 38}$$

Done.

13.3 Challenges

1. Let m, k, and n be 3 positive integers satisfying $m \le k \le n$. Show that

$$C_n^k C_k^m = C_n^m C_{n-m}^{k-m} = C_n^{k-m} C_{n-k+m}^k \qquad (13.3)$$

2. Show that

$$\frac{1}{C_n^k} = \frac{k}{k-1}\left(\frac{1}{C_{n-1}^{k-1}} - \frac{1}{C_n^{k-1}}\right) \qquad (13.4)$$

3. Simplify

$$\sum_{k=1}^{n} k^2 C_n^k$$

4. Let positive integers m and n satisfy $m \le n$. Prove

$$\sum_{k=m}^{n} C_n^k C_k^m = 2^{n-m} C_n^m$$

5. Show that

$$C_n^1 - \frac{1}{2}C_n^2 + \frac{1}{3}C_n^3 - \cdots + (-1)^{n+1}C_n^n = 1 + \frac{1}{2} + \frac{1}{3} + \cdots + \frac{1}{n}$$

6. Show that

$$\frac{1}{C_{2n}^0} - \frac{2}{C_{2n}^1} + \cdots - \frac{2n}{C_{2n}^{2n-1}} = 0$$

7. Evaluate

$$\sum_{n=2016}^{\infty} \frac{1}{C_n^{2016}}$$

Lecture 14

$$\cos \frac{\pi}{2n+1} \cos \frac{2\pi}{2n+1} \cdots \cos \frac{2^n \pi}{2n+1}$$

14.1 Objectives

- Simplify trigonometric expressions using trigonometric identities

14.2 Examples

Evaluating trigonometric expressions will be tested in intermediate to advanced level competitions. These problems may appear on their own or as parts of other problems. For example, some geometry problems can be solved by the trigonometry method. [1]

The method of choice to simplify and evaluate a complex trigonometric expression is to apply appropriate trigonometric identities. There are many trigonometric identities available. Hence, the key to success is to choose appropriate ones.

---

[1]Solving geometry problems using trigonometry is discussed in the book *Geometry Technique* by the same author.

Let's consider the following example.

Example 14.2.1

Compute the value of $\cos \frac{2\pi}{7} \cos \frac{4\pi}{7} \cos \frac{8\pi}{7}$.

Solution

Multiplying the given expression by $\sin \frac{2\pi}{7}$ yields:

$$\sin \frac{2\pi}{7} \cos \frac{2\pi}{7} \cos \frac{4\pi}{7} \cos \frac{8\pi}{7}$$
$$= \frac{1}{2} \sin \frac{4\pi}{7} \cos \frac{4\pi}{7} \cos \frac{8\pi}{7}$$
$$= \frac{1}{2^2} \sin \frac{8\pi}{7} \cos \frac{8\pi}{7}$$
$$= \frac{1}{2^3} \sin \frac{16\pi}{7}$$
$$= \frac{1}{8} \sin \frac{2\pi}{7}$$

$$\therefore \quad \sin \frac{2\pi}{7} \cos \frac{2\pi}{7} \cos \frac{4\pi}{7} \cos \frac{8\pi}{7} = \frac{1}{8} \sin \frac{2\pi}{7}$$

$$\implies \quad \cos \frac{2\pi}{7} \cos \frac{4\pi}{7} \cos \frac{8\pi}{7} = \boxed{\frac{1}{8}}$$

Done.

Careful observation often plays a critical role in choosing the most appropriate identities in order to solve problems at hand. For instance, angles involved in *Example 14.2.1* form a geometric cosine sequence whose common ratio is 2. Multiplying them with an appropriate sine function will construct a self-recurring sequence. This is a strong hint to use the double angle formula.

14.3 Challenges

1) Find the value of $\cos 20° \cos 40° \cos 80°$.

2) Let $\alpha \in \left(\frac{3\pi}{2}, 2\pi \right)$. Simplify

$$\sqrt{\frac{1}{2} - \frac{1}{2}\sqrt{\frac{1}{2} + \frac{1}{2} \cdot \cos 2\alpha}}$$

3) Simplify $\cos x \cos 2x \cdots \cos 2^{n-1}x$.

4) Show that

$$\sin x + 2 \sin 2x + \cdots + n \sin nx = \frac{(n+1)\sin nx - n \sin(n+1)x}{2(1 - \cos x)}$$

5) Compute the value of

$$\cos \frac{\pi}{2n+1} \cdot \cos \frac{2\pi}{2n+1} \cos \frac{3\pi}{2n+1} \cdots \cos \frac{n\pi}{2n+1}$$

6) Evaluate

$$\cos \frac{2\pi}{2n+1} + \cos \frac{4\pi}{2n+1} + \cdots + \cos \frac{2n\pi}{2n+1}$$

7) Evaluate

$$\cos \frac{\pi}{2n+1} + \cos \frac{3\pi}{2n+1} + \cdots + \cos \frac{(2n-1)\pi}{2n+1}$$

8) Without using a calculator, find the value of

$$\cos \frac{\pi}{13} + \cos \frac{3\pi}{13} + \cos \frac{9\pi}{13}$$

9) Evaluate

$$(1 + \tan 1°)(1 + \tan 2°) \cdots (1 + \tan 44°)(1 + \tan 45°)$$

10) Evaluate

$$\cos \frac{\pi}{2n+1} \cdot \cos \frac{2\pi}{2n+1} \cdot \cos \frac{3\pi}{2n+1} \cdots \cos \frac{n\pi}{2n+1}$$

11) Let sequence $\{a_n\}$ satisfy the condition: $a_1 = \frac{\pi}{6}$ and $a_{n+1} = \arctan(\sec a_n)$, where $n \in Z^+$. There exists a positive integer m such that $\sin a_1 \cdot \sin a_2 \cdots \sin a_m = \frac{1}{100}$. Find m.

(Ref 2014 China)

Lecture 15

$$\sin\frac{\pi}{n}\sin\frac{2\pi}{n}\cdots\sin\frac{(n-1)\pi}{n}$$

15.1 Objectives

- Evaluate trigonometric expressions using complex number
- Express sin and cos functions using complex number

15.2 Examples

Complex number is a powerful tool to solve many types of mathematical problems including trigonometry. This will be investigated in this lecture.

There are several different ways to solve trigonometric problems using complex numbers. One way is to utilize unit roots. This is illustrated in the following example.

Example 15.2.1

Let integer $n \geq 2$, prove that

$$\sin \frac{\pi}{n} \cdot \sin \frac{2\pi}{n} \cdots \sin \frac{(n-1)\pi}{n} = \frac{n}{2^{n-1}}$$

Solution

Let $Z = e^{i\frac{2\pi}{n}}$. Then $Z, Z^2, \cdots, Z^{n-1}$ are $(n-1)$ complex roots of the equation $x^n = 1$. Because none of them equals 1, they must be the roots of the equation

$$x^{n-1} + x^{n-2} + \cdots + x + 1 = 0$$

$$\therefore \quad x^{n-1} + x^{n-2} + \cdots + x + 1 = (x - Z)(x - Z^2) \cdots (x - Z^{n-1})$$

Letting $x = 1 \implies n = (1 - Z)(1 - Z^2) \cdots (1 - Z^{n-1})$. Taking modulus on both sides yields:

$$n = |1 - Z| \cdot |1 - Z^2| \cdots |1 - Z^{n-1}| \qquad (15.1)$$

On the other hand, any term on the right can be computed as

$$|1 - Z^k| = |1 - (\cos \frac{2k\pi}{n} + i \sin \frac{2k\pi}{n})| = 2 \sin \frac{k\pi}{n}$$

Setting this to *(15.1)*:

$$n = 2 \sin \frac{\pi}{n} \cdot 2 \sin \frac{2\pi}{n} \cdots 2 \sin \frac{(n-1)\pi}{n}$$

$$n = 2^{n-1} \left(\sin \frac{\pi}{n} \cdot \sin \frac{2\pi}{n} \cdots \sin \frac{(n-1)\pi}{n} \right)$$

$$\therefore \quad \sin \frac{\pi}{n} \cdot \sin \frac{2\pi}{n} \cdots \sin \frac{(n-1)\pi}{n} = \frac{n}{2^{n-1}}$$

Done.

Another common technique is to convert trigonometric expressions to complex polynomials. Let $z = \cos\theta + i\sin\theta$, then $\frac{1}{z} = \bar{z} = \cos\theta - i\sin\theta$. It follows that

$$\cos\theta = \frac{1}{2} \times \left(z + \frac{1}{z}\right) \quad \text{and} \quad \sin\theta = -\frac{1}{2} \times \left(z - \frac{1}{z}\right)i \quad (15.2)$$

By replacing sin and cos with z and $\bar{z}$, a trigonometric expression can be converted to a polynomial with respect to z and $\bar{z}$. Such polynomials are usually easier to handle than trigonometric expressions. Some practice problems require this technique.

15.3 Challenges

1) Use *Equation 15.2* on *page 65* to prove $\sin^2\theta + \cos^2\theta = 1$.

2) Let sequences a_n and b_n satisfy $a_n = a_{n-1}\cos\theta - b_{n-1}\sin\theta$ and $b_n = a_{n-1}\sin\theta + b_{n-1}\cos\theta$. If $a_1 = 1$ and $b_1 = \tan\theta$, where θ is a known real number, find the general formulas for a_n and b_n

3) Simplify
$$\sin x + \sin 2x + \cdots + \sin nx$$
and
$$\cos x + \cos 2x + \cdots + \cos nx$$

4) Solve the equation $\cos\theta + \cos 2\theta + \cos 3\theta = \sin\theta + \sin 2\theta + \sin 3\theta$.

5) Solve the equation $\cos^2 x + \cos^2 2x + \cos^2 3x = 1$ in $(0, 2\pi)$.
   (Ref 1962 IMO)

6) Let A, B, and C be a triangle's three inner angles. Let $a = \cos A + i\sin A$, $b = \cos B + i\sin B$, and $c = \cos C + i\sin C$. Show that
$$abc = -1 \quad (15.3)$$

7) Let A, B, and C be angles of a triangle. If $\cos 3A + \cos 3B + \cos 3C = 1$, determine the largest angle of the triangle.

8) Evaluate $\sin \theta + \frac{1}{2} \cdot \sin 2\theta + \frac{1}{4} \cdot \sin 3\theta + \cdots$.

9) Prove for every positive integer n and real number $x \neq \frac{k\pi}{2^t}$ where $t = 0, 1, 2, \cdots$ and k is an integer, the following relation always holds:

$$\frac{1}{\sin 2x} + \frac{1}{\sin 4x} + \cdots + \frac{1}{\sin 2^n x} = \frac{1}{\tan x} - \frac{1}{\tan 2^n x}$$

(Ref 1966 IMO)

10) Let S_n be the minimal value of $\displaystyle\sum_{k=1}^{n} \sqrt{a_k^2 + b_k^2}$ where $\{a_k\}$ is an arithmetic sequence whose first term is 4 and common difference is 8. $b_1, b_2, \cdots, b_n$ are positive real numbers satisfying $\displaystyle\sum_{k=1}^{n} b_k = 17$. If there exist a positive integer n such that S_n is also an integer, find n.

Complex numbers can also be used to solve other types of problems. For this problem, the following property can be used:

$$|z_1 + z_2 + \cdots + z_n| \leq |z_1| + |z_2| + | + \cdots + |z_n|$$

Solutions

Lecture 1

1) Compute: $1 + 3 + 5 + \cdots + (2n - 1)$.

$1 + 3 + 5 + \cdots + (2n - 1) = \frac{(1+(2n-1)) \cdot n}{2} = n^2$.

---

2) Use the diagram below to explain the sum of first n odd integers is a perfect square.

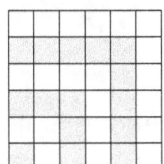

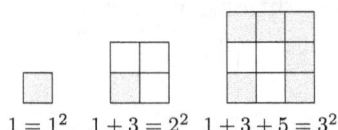

$1 = 1^2 \quad 1 + 3 = 2^2 \quad 1 + 3 + 5 = 3^2$

3) Suppose no term in an arithmetic sequence $\{a_n\}$ equals 0. Let S_n be the sum of its first n terms. If $S_{2n-1} = a_n^2$, express its nth term a_n with respect to n.

By *Theorem 1.2.3* on *page 4*, we have

$$S_{2n-1} = a_n \cdot n \implies a_n^2 = a_n(2n-1) \implies \boxed{a_n = 2n - 1} \quad (\because a_n \neq 0)$$

> ⓘ *Tip: $S_{2n-1} = n^2$ is a mathematical way to say the sum of first $2n - 1$ terms of a sequence is a perfect square.*

4) The sum of n different positive integers is less than 100. What is the greatest possible value for n?

In order to have as many numbers as possible, all these numbers should be as small as possible. n distinct smallest possible numbers are $1, 2, \cdots, n$. Because their sum is less than 100, it must hold that

$$\frac{n(n+1)}{2} < 100 \implies n \leq 13$$

Therefore the answer is $\boxed{13}$.

5) Determine all pairs (a, b) of real numbers such that $10, a, b, ab$ is an arithmetic progression.

From the given conditions, we have

$$\begin{cases} 10 + b &= 2a \\ a + ab &= 2b \end{cases}$$

Solving this system yields $(a, b) = \boxed{(4, -2), (\frac{5}{2}, -5)}$.

6) A sequence $\{a_n\}$ satisfies $a_n + a_m = a_{n+m}$ for any positive integers n and m. If $a_1 = \frac{1}{2013}$, find the sum of its first 2013 terms.

(Ref 2013 China)

$$
\begin{aligned}
a_2 &= a_1 + a_1 = 2a_1 \\
a_3 &= a_2 + a_1 = 3a_1 \\
&\cdots \\
a_{2013} &= a_{2012} + a_1 = 2013a_1
\end{aligned}
$$

$$
\begin{aligned}
\therefore \quad & a_1 + a_2 + \cdots + a_{2013} \\
=& a_1 + 2a_1 + \cdots 2013a_1 \\
=& (1 + 2 + \cdots + 2013)a_1 \\
=& \frac{(1 + 2013) \times 2013}{2} \times \frac{1}{2013} \\
=& \boxed{1007}
\end{aligned}
$$

7) Let S_n be the sum of first n terms of an arithmetic sequence. Show that S_n must be in the form of $An^2 + Bn$ where A and B are two constants.

Let the 1st and n^{th} terms of this arithmetic sequence be a_1 and a_n, respectively. Then $a_n = a_1 + (n-1)d$ where d is the common difference. It follows that

$$
S_n = \frac{(a_1 + a_n)n}{2} = \frac{(a_1 + a_1 + (n-1)d)n}{2} = \frac{d}{2} \cdot n^2 + \left(a_1 - \frac{d}{2}\right) \cdot n
$$

Hence the claim holds where both $A = \frac{d}{2}$ and $B = \left(a_1 - \frac{d}{2}\right)$ are constants for a given arithmetic sequence.

8) Let the sum of first n terms of arithmetic sequence $\{a_n\}$ be S_n, and the sum of first n terms of arithmetic sequence $\{b_n\}$ be T_n. If $S_n : T_n = 2n : 3n + 7$, compute the value of $a_8 : b_6$.

(Ref 2013 China)

The sum of n consecutive terms in an arithmetic sequence must be in the form of $An^2 + Bn$ where A and B are two constants.

Let $S_n = A_1 n^2 + B_1$ and $T_n = A_2 n^2 + B_2$, when A_1, B_1, A_2, and B_2 are to be determined constant coefficients. Therefore

$$\frac{S_n}{T_n} = \frac{A_1 n^2 + B_1 n}{A_2 n^2 + B_2 n} = \frac{A_1 n + B_1}{A_2 n + B_2} = \frac{2n}{3n + 7}$$

It follows that $B_1 = 0$ and $A_1 : A_2 : B_2 = 2 : 3 : 7$. Let $A_1 = 2k, A_2 = 3k$, and $B_2 = 7k$ where k is a constant.

$$S_n = 2kn^2 \implies a_8 = S_8 - S_7 = 128k - 98k = 30k$$

$$T_n = 3kn^2 + 7kn \implies b_6 = T_6 - T_5 = 40k$$

$$\therefore \quad \frac{a_8}{b_6} = \frac{30k}{40k} = \boxed{\frac{3}{4}}$$

9) If the coefficients of the 5th, 6th and 7th terms in the expanded form of $(x^{-\frac{4}{3}} + x)^n$ form an arithmetic sequence, find the constant term in the expanded form.

The coefficients of the $5^{th}, 6^{th}$, and 7th terms ar C_n^4, C_n^5, and C_n^6, respectively. Therefore

$$2C_n^5 = C_n^4 + C_n^6$$

$$2 \times \frac{n!}{5!(n-5)!} = \frac{n!}{4!(n-4)!} + \frac{n!}{6!(n-6)!}$$

$$\frac{2}{5(n-5)} = \frac{1}{(n-4)(n-5)} + \frac{1}{6 \times 5}$$

$$n^2 - 21n + 98 = 0$$

$$n_{1,2} = 7, 14$$

Let the constant term in the expanded form be

$$C_n^k (x^{-\frac{4}{3}})^k x^{n-k} = C_n^k x^{n-\frac{7}{3}k} \implies k = \frac{3}{7}n$$

When $n = 7 \implies k = 3$, its constant term is $C_7^3 = \boxed{35}$.

When $n = 14 \implies k = 6$, its constant term is $C_{14}^6 = \boxed{2003}$.

10) Given a sequence $\{a_n\}$, if $a_n \neq 0$, $a_1 = 1$, and $3a_n a_{n-1} + a_n - a_{n-1} = 0$ for any $n \geq 2$, find the general term of a_n.

$$3a_n a_{n-1} + a_n - a_{n-1} = 0 \implies a_{n-1} - a_n = 3a_n a_{n-1}$$

Because all terms are non zero, dividing both sides of the above equation by $a_n a_{n-1}$ yields

$$\frac{1}{a_n} - \frac{1}{a_{n-1}} = 3$$

This implies $\{\frac{1}{a_n}\}$ is an arithmetic sequence whose first term is $\frac{1}{a_1} = 1$ and common difference is 3. It follows that

$$\frac{1}{a_n} = 1 + 3 \times (n-1) = 3n - 2 \implies \boxed{a_n = \frac{1}{3n-2}}$$

ⓘ   *Tip: Quite often, a complex relationship can be simplified by polynomial transformation.*

Lecture 2

1) Is it possible for a sequence to be both arithmetic and geometric?

Yes. A sequence of constants is both arithmetic (common difference is zero) and geometric (common ratio is one).

2) Simplify the expression: $1 + x + x^2 + \cdots + x^n$.

If $x = 1$, the original expression equals n. Otherwise, by *Formula 2.2.1* on *page 8*, where $a_1 = 1$ and $r = x$:

$$1 + x + x^2 + \cdots + x^n = \frac{1 - x^n}{1 - x}$$

 Caution: Do not forget to discuss the case when the common ratio equals 1 before applying the geometric sequence sum formula.

3) Explain the following polynomial factorization identity using the concept of geometric sequence:

$$1 - x^n = (1 - x)(1 + x + x^2 + \cdots + x^{n-1})$$

This important polynomial identity can be directly obtained from previous practice's result.

4) Compute $2 + 2^3 + 2^5 + \cdots + 2^{2n-1}$.

This is a n-term geometric sequence whose common ratio is 2^2. Therefore its sum is given by:

$$= 2 \times \frac{1 - (2^2)^n}{1 - 2^2} = \frac{2}{3} \times (2^{2n} - 1)$$

5) Let $c_1, c_2, c_3, \cdots$ be a series of concentric circles whose radii form a geometric sequence with common ratio as q. Suppose the areas of rings which are formed by two adjacent circles are $S_1, S_2, S_3, \cdots$. Which statement below is correct regarding the sequence $\{S_n\}$?

(A) It is not a geometric sequence

(B) It is a geometric sequence and its common ratio is q

(C) It is a geometric sequence and its common ratio is q^2

(D) It is a geometric sequence and its common ratio is $q^2 - 1$

The answer is (C).

Suppose the radius of the smallest circle is r, then the radii of the kth and $(k+1)^{th}$ are rq^{k-1} and rq^k, respectively. Hence:

$$\frac{S_k}{S_{k-1}} = \frac{\pi(rq^k)^2 - \pi(rq^{k-1})^2}{\pi(rq^{k-1})^2 - \pi(rq^{k-2})^2} = \frac{q^{2k} - q^{2k-2}}{q^{2k-2} - q^{2k-4}} = q^2$$

ⓘ   *Tip: In order to prove a sequence is geometric, it is sufficient to show that the ratio between two adjacent terms is a constant.*

6) Suppose all the terms in a geometric sequence $\{a_n\}$ are positive. If $|a_2 - a_3| = 14$ and $|a_1 a_2 a_3| = 343$, find a_5.

Firstly, $a_1 a_2 a_3 = a_2^3 \implies a_2^3 = 343 \implies a_2 = 7$.

Secondly, $|a_2 - a_3| = 14 \implies |7 - a_3| = 14 \implies a_3 = 21$ ($\because$ $a_3 > 0$).

Therefore the common ratio of this sequence equal $a_3/a_2 = 3$. It follows $a_5 = a_3 \times 3^2 = \boxed{189}$.

7) Let sequence $\{a_n\}$ satisfy $a_0 = 1$ and $a_n = \frac{\sqrt{1+a_{n-1}^2}-1}{a_{n-1}}$. Prove $a_n > \frac{\pi}{2^{n+2}}$.

(Ref 1990 Hungarian)

It is easy to see that $a_n > 0$. Let $a_n = \tan \alpha_n$, then $\alpha \in \left(0, \frac{\pi}{2}\right)$.

$$a_n = \frac{\sqrt{1 + \tan^2 \alpha_{n-1}} - 1}{\tan \alpha_{n-1}} = \frac{1 - \cos \alpha_{n-1}}{\sin \alpha_{n-1}} = \tan \frac{\alpha_{n-1}}{2}$$

It follows

$$\tan \alpha_n = \tan \frac{\alpha_{n-1}}{2} \implies \alpha_n = \frac{\alpha_{n-1}}{2}$$

$$a_0 = 1 \implies \alpha_0 = \frac{\pi}{4}$$

Hence, $\{\alpha_n\}$ is a geometric sequence whose initial term $\alpha_0 = \frac{\pi}{4}$ and common ratio is $\frac{1}{2}$. This implies

$$\alpha_n = \left(\frac{1}{2}\right)^n \cdot \frac{\pi}{4} = \frac{\pi}{2^{n+2}}$$

Given $\alpha \in \left(0, \frac{\pi}{2}\right)$, we have

$$a_n = \tan \alpha_n > \alpha_n = \frac{\pi}{2^{n+2}}$$

Lecture 3

1) A sequence satisfies $a_1 = 3, a_2 = 5$, and $a_{n+2} = a_{n+1} - a_n$ for $n \geq 1$ What is the value of a_{2016}?

By trying a few numbers, we find the sequence repeats every 6 terms:

$$3, 5, 2, -3, -5, -2, 3, 5, \cdots$$

Because $6 \mid 2016$, we conclude that $a_{2016} = a_6 = \boxed{-2}$.

2) The sum of the first n terms of sequence $\{a_n\}$ is given by the formula $S_n = n^2 + n + 3$ What is the value of a_{10}?

$$a_{10} = S_{10} - S_9 = (10^2 + 10 + 3) - (9^2 + 9 + 3) = 20$$

3) Suppose every term in the sequence

$$1, 2, 1, 2, 2, 2, 1, 2, 2, 2, 2, 2, 1, \cdots$$

is either 1 or 2. If there are exactly $(2k-1)$ *twos* between the kth *one* and the $(k+1)^{th}$ *one*, find the sum of its first 2014 terms.
(Ref 2014 China)

Let's divide this sequence into a series of groups. Each group starts with 1 and is followed by as many 2s as possible:

$$\{1, 2\}, \quad \{1, 2, 2, 2\}, \quad \{1, 2, 2, 2, 2, 2\}, \quad \{1, \cdots$$

It is clear that the kth group contains $2k$ terms whose sum equals $1 + (2k - 1) \times 2 = 4k - 1$.

The numbers of terms in these groups form an arithmetic sequence. The total number of terms in first n groups equal

$$2 \times 1 + 2 \times 2 + \cdots + 2 \times n = 2 \times \frac{n(n + 1)}{2} = n \times (n + 1)$$

$$n \times (n + 1) \leq 2014 \implies n = 44$$

This means that 2014 terms in the original sequence cover 44 complete group and 1 partial group. Hence there are 45 *ones* in the first 2014 terms and $(2014 - 45)$ *twos*. Therefore, their sum equal

$$45 \times 1 + (2014 - 45) \times 2 = \boxed{3983}$$

4) Simplify $\frac{1}{2} + \frac{2}{2^2} + \frac{3}{2^3} + \cdots + \frac{n}{2^n}$

This problem is a special case of *Example 3.2.1* on *page 11* where $x = \frac{1}{2}$. Therefore the answer is

$$\frac{(1/2)(1 - (1/2)^n)}{(1 - 1/2)^2} - \frac{n(1/2)^{n+1}}{1 - 1/2} = 2 - \frac{n + 2}{2^n}$$

5) Let sequence $\{a_n\}$ satisfy $a_1 = 2$ and $a_{n+1} = \frac{2(n+2)}{n+1} a_n$ where $n \in \mathbb{Z}^+$. Compute the value of

$$\frac{a_{2014}}{a_1 + a_2 + \cdots + a_{2013}}$$

(Ref 2014 China)

The given conditions lead to:

$$a_n = \frac{2(n+1)}{n} \cdot a_{n-1}$$
$$= \frac{2(n+1)}{n} \cdot \frac{2n}{n-1} \cdot a_{n-2}$$
$$= \frac{2(n+1)}{n} \cdot \frac{2n}{n-1} \cdot \frac{2(n-1)}{n-2} \cdot a_{n-3}$$
$$= \cdots$$
$$= \frac{2(n+1)}{n} \cdot \frac{2n}{n-1} \cdot \frac{2(n-1)}{n-2} \cdots \frac{2 \times 3}{2} \cdot a_1$$
$$= 2^{n-1}(n+1)$$

Let S_n be the sum of first n terms of this sequence. Then

$$S_n = 2 + 2 \times 3 + 2^2 \times 4 + \cdots + 2^{n-1}(n+1)$$
$$2S_n = 2 \times 2 + 2^2 \times 3 + 2^3 \times 4 + \cdots + 2^n(n+1)$$

Subtracting the 1st equation from the 2nd one yields:

$$S_n = 2^n(n+1) - (2^{n-1} + 2^{n-2} + \cdots + 2 + 2) = 2^n(n+1) - 2^n = 2^n n$$

Hence,

$$\frac{a_{2014}}{a_1 + a_2 + \cdots + a_{2013}} = \frac{2^{2013} \times (2014+1)}{2^{2013} \times 2013} = \boxed{\frac{2015}{2013}}$$

ⓘ   *Tip: Have you noticed that $a_n = 2^{n-1}(n+1)$ defines a similar sequence as the one in Example 3.2.1 on page 11?*

6) Let α and β be the two roots of the equation $x^2 - x - 1 = 0$. If

$$a_n = \frac{\alpha^n - \beta^n}{\alpha - \beta} \quad (n = 1, 2, \cdots)$$

Show that, for any positive integer n, it always hold

$$a_{n+2} = a_{n+1} + a_n$$

Because α and β are the two roots of $x^2 - x - 1 = 0$, we have

$$\alpha^2 = \alpha + 1 \implies \alpha^{n+2} = \alpha^{n+1} + \alpha^n$$

$$\beta^2 = \beta + 1 \implies \beta^{n+2} = \beta^{n+1} + \beta^n$$

$$\therefore \quad a_{n+2} = \frac{\alpha^{n+2} - \beta^{n+2}}{\alpha - \beta}$$

$$= \frac{(\alpha^{n+1} + \alpha^n) - (\beta^{n+1} + \beta^n)}{\alpha - \beta}$$

$$= \frac{(\alpha^{n+1} - \beta^{n+1}) + (\alpha^n - \beta^n)}{\alpha - \beta}$$

$$= \frac{(\alpha^{n+1} - \beta^{n+1})}{\alpha - \beta} + \frac{(\alpha^n - \beta^n)}{\alpha - \beta}$$

$$= a_{n+1} + a_n$$

7) Let sequence $\{b_n\}$ satisfy $2b_{n+1} = b_n + 3$ and $b_1 = 5$, express b_n with respect to n.

From the given conditions, it is easy to derive by polynomial transformation that

$$2b_{n+1} = b_n + 3 \implies (b_{n+1} - 3) = \frac{1}{2} \times (b_n - 3)$$

This means $\{b_n - 3\}$ forms a geometric sequence. Its first term is $5 - 3 = 2$ and its common ratio is $\frac{1}{2}$. Hence

$$b_n - 3 = 2 \times \left(\frac{1}{2}\right)^{n-1} = \frac{1}{2^{n-2}} \implies \boxed{b_n = 3 + \frac{1}{2^{n-2}}}$$

ⓘ *Tip: A linear recurrence in the form of $a_n = pa_{n-1} + q$ can be transformed to the form $a_n + \alpha = \beta(a_{n-1} + \alpha)$ where p, q, α and β are all constants.*

8) If a sequence $\{a_n\}$ satisfies $a_1 = 1$ and

$$a_{n+1} = \frac{1}{16}\left(1 + 4a_n + \sqrt{1 + 24a_n}\right)$$

, express a_n in terms of n.

(Ref 1981 IMO Shortlist)

Construct a new sequence $\{b_n\}$ such that $b_n = \sqrt{1 + 24a_n}$. It is apparent that $b_1 = 5$ and every term in $\{b_n\}$ is non-negative.

$$b_n^2 = 1 + 24a_n \implies a_n = \frac{b_n^2 - 1}{24}$$

Then from the given recurrence, we have

$$\frac{b_{n+1}^2 - 1}{24} = \frac{1}{16} \times \left(1 + 4 \times \frac{b_n^2 - 1}{24} + b_n\right)$$

This above relationship can be simplified to

$$(2b_{n+1})^2 = (b_n + 3)^2 \implies 2b_{n+1} = b_n + 3 \qquad (\because b_n \geq 0)$$

By the previous practice, we have

$$2b_{n+1} = b_n + 3 \implies b_n = 3 + \frac{1}{2^{n-2}} = 3 + 2^{2-n}$$

$$\therefore \quad a_n = \frac{b_n^2 - 1}{24} = \frac{2^{2n-1} + 3 \times 2^{n-1} + 1}{3 \times 2^{2n-1}}$$

Lecture 4

1) Convert $0.\dot{7}$ to a fraction.

$$0.\overline{7} = 0.7 + 0.07 + 0.007 + \cdots$$
$$= \frac{7}{10} + \frac{7}{100} + \frac{7}{1000} + \cdots$$
$$= \frac{\frac{7}{10}}{1 - \frac{1}{10}}$$
$$= \frac{7}{9}$$

2) Convert $0.\dot{7}\dot{3}$ to a fraction.

$$0.\overline{73} = 0.73 + 0.0073 + 0.000073 + \cdots$$
$$= \frac{73}{100} + \frac{73}{10000} + \frac{73}{1000000} + \cdots$$
$$= \frac{\frac{73}{100}}{1 - \frac{1}{100}}$$
$$= \frac{73}{99}$$

A recurring decimal can be converted to a fraction. Its numerator is the repetend. Its denominator is in the form of $9 \cdots 9$ whose length equals the repetend's length.

3) Let

$$f(r) = \sum_{j=2}^{2016} \frac{1}{j^r} = \frac{1}{2^r} + \frac{1}{3^r} + \cdots + \frac{1}{2016^r}$$

Find

$$\sum_{k=2}^{\infty} f(k)$$

First, expand $\sum_{k=2}^{\infty} f(k)$:

$$f(2) = \frac{1}{2^2} + \frac{1}{3^2} + \frac{1}{4^2} + \cdots + \frac{1}{2016^2}$$

$$f(3) = \frac{1}{2^3} + \frac{1}{3^3} + \frac{1}{4^3} + \cdots + \frac{1}{2016^3}$$

$$f(4) = \frac{1}{2^4} + \frac{1}{3^4} + \frac{1}{4^4} + \cdots + \frac{1}{2016^4}$$

$$\cdots$$

Next, add these terms by column. Each forms an infinite geometric sequence.

$$\frac{1}{2^2} + \frac{1}{2^3} + \frac{1}{2^4} + \cdots = \frac{1}{2^2} \times \frac{1}{1-\frac{1}{2}} = \frac{1}{1 \times 2}$$

$$\frac{1}{3^2} + \frac{1}{3^3} + \frac{1}{3^4} + \cdots = \frac{1}{3^2} \times \frac{1}{1-\frac{1}{3}} = \frac{1}{2 \times 3}$$

$$\frac{1}{4^2} + \frac{1}{4^3} + \frac{1}{4^4} + \cdots = \frac{1}{4^2} \times \frac{1}{1-\frac{1}{4}} = \frac{1}{3 \times 4}$$

$$\cdots$$

$$\frac{1}{2016^2} + \frac{1}{2016^3} + \frac{1}{2016^4} + \cdots = \frac{1}{2016^2} \times \frac{1}{1-\frac{1}{2016}} = \frac{1}{2015 \times 2016}$$

Therefore, the sum equals[1]

$$\frac{1}{1 \times 2} + \frac{1}{2 \times 3} + \frac{1}{3 \times 4} + \cdots + \frac{1}{2015 \times 2016}$$

$$= \left(\frac{1}{1} - \frac{1}{2}\right) + \left(\frac{1}{2} - \frac{1}{3}\right) + \cdots + \left(\frac{1}{2015} - \frac{1}{2016}\right)$$

$$= \boxed{\frac{2015}{2016}}$$

---

[1]Evaluating expressions in such form will be discussed in *Lecture 7*.

Solution

4) Find all x such that $\displaystyle\sum_{k=1}^{\infty} kx^k = 20$.

$$\sum_{k=1}^{\infty} kx^k = x + 2x^2 + 3x^3 + \cdots$$

$$= (x + x^2 + x^3 + \cdots)(1 + x + x^2 + \cdots)$$

$$= \left(x \cdot \frac{1}{1-x}\right)\left(\frac{1}{1-x}\right) \qquad \text{(if } |x| < 1\text{)}$$

Therefore

$$\frac{x}{(1-x)^2} = 20 \implies x_{1,2} = \frac{5}{4} \quad \text{or} \quad \frac{4}{5} \implies \boxed{x = \frac{4}{5}} \quad (\because |x| < 1)$$

5) Let $a_1, a_2, \cdots, a_n > 0, n \geq 2$, and $a_1 + a_2 + \cdots + a_n = 1$. Prove

$$\frac{a_1}{2 - a_1} + \frac{a_2}{2 - a_2} + \cdots + \frac{a_n}{2 - a_n} \geq \frac{n}{2n - 1}$$

(Ref 1984 Bulkans) Tip: write each term as the sum of an infinite geometric sequence and then apply the following inequality.

$$\frac{a_1 + a_2 + \cdots + a_n}{n} \leq \sqrt[m]{\frac{a_1^m + a_2^m + \cdots + a_n^m}{n}}$$

Because $0 < a_1 < 1$, it must hold that $\left|\frac{a_1}{2}\right| < 1$. Therefore

$$\frac{a_1}{2 - a_1} = \frac{\frac{a_1}{2}}{1 - \frac{a_1}{2}} = \frac{a_1}{2} + \left(\frac{a_1}{2}\right)^2 + \left(\frac{a_1}{2}\right)^3 + \cdots + \left(\frac{a_1}{2}\right)^n + \cdots$$

It follows that

$$\frac{a_1}{2-a_1} + \frac{a_2}{2-a_2} + \cdots + \frac{a_n}{2-a_n} \geq \frac{n}{2n-1}$$

$$= \left(\frac{a_1}{2} + \frac{a_2}{2} + \cdots + \frac{a_n}{2}\right) +$$

$$\left(\left(\frac{a_1}{2}\right)^2 + \left(\frac{a_2}{2}\right)^2 + \cdots + \left(\frac{a_n}{2}\right)^2\right) +$$

$$\cdots$$

$$\left(\left(\frac{a_1}{2}\right)^n + \left(\frac{a_2}{2}\right)^n + \cdots + \left(\frac{a_n}{2}\right)^n\right) +$$

$$\cdots$$

$$\geq \frac{1}{2} + \frac{\left(\frac{a_1}{2} + \frac{a_2}{2} + \cdots + \frac{a_n}{2}\right)^2}{n} + \frac{\left(\frac{a_1}{2} + \frac{a_2}{2} + \cdots + \frac{a_n}{2}\right)^3}{n^2}$$

$$+ \cdots + \frac{\left(\frac{a_1}{2} + \frac{a_2}{2} + \cdots + \frac{a_n}{2}\right)^n}{n^{n-1}} + \cdots$$

$$= \frac{1}{2} + \frac{1}{2^2 n} + \frac{1}{2^3 n^2} + \cdots + \frac{1}{2^n n^{n-1}} + \cdots$$

$$= \frac{\frac{1}{2}}{1 - \frac{1}{2n}} = \frac{n}{2n-1}$$

Lecture 5

1) Evaluate $\sqrt{7 + \sqrt{2 + \sqrt{2 + \sqrt{2 + \cdots}}}}$.

The difference between this problem and *Example 5.2.1* is that this expression has different terms. Consequently, the equation method cannot be applied directly. That being said, it is possible to first evaluate the inside part and then compute the rest.

$$\because \quad \sqrt{2 + \sqrt{2 + \sqrt{2 + \cdots}}} = 2 \qquad \textit{Example 5.2.1 on page 19}$$

$$\therefore \quad \sqrt{7 + \sqrt{2 + \sqrt{2 + \sqrt{2 + \cdots}}}} = \sqrt{7 + 2} = 3$$

2) Simplify

$$\cfrac{1}{2 + \cfrac{1}{2 + \cdots}}$$

Let the result be S.

$$S = \cfrac{1}{2 + \cfrac{1}{2 + \cdots}} \implies S = \frac{1}{2 + S} \implies S^2 + 2S - 1 = 0$$

It is obvious that $S > 0$. Solving this equation leads

$$S = \boxed{\sqrt{2} - 1}$$

.

3) Compute

$$\cfrac{1}{\cfrac{1}{\frac{1}{\cdots} + 1 + \frac{1}{\cdots}} + 1 + \cfrac{1}{\frac{1}{\cdots} + 1 + \frac{1}{\cdots}}}$$

Suppose the result is S. Then $S = \frac{1}{S+1+S} \implies 2S^2 + S - 1 = 0$. This equation has only one positive solution $S = \frac{1}{2}$. Therefore the answer is $\boxed{\dfrac{1}{2}}$.

4) Simplify

$$(\sqrt{2})^{(\sqrt{2})^{(\sqrt{2})^{\cdots}}}$$

Suppose the result is S. Then

$$S = (\sqrt{2})^{(\sqrt{2})^{(\sqrt{2})^{\cdots}}} \implies S = (\sqrt{2})^S \implies S^2 = 2^S$$

Obviously, 2 is one solution. Hence the answer is $\boxed{2}$.

5) Compute

$$\sqrt[3]{6 + \sqrt[3]{6 + \sqrt[3]{6 + \cdots}}}$$

Let $x = \sqrt[3]{6 + \sqrt[3]{6 + \sqrt[3]{6 + \cdots}}} \implies x^3 = 6 + x$. It is a cubic equation. By guessing and trying, we find $x = 2$ is one solution[2]. Hence the original expression equals $\boxed{2}$.

---

[2]Similar to quadratic formulas, solutions to cubic equations are available though they are more complex.

6) Let x be a positive number, use two different reasonings to show that the following relation holds:

$$\sqrt{x\sqrt{x\sqrt{x\sqrt{\cdots}}}} = x$$

1) Let $S = \sqrt{x\sqrt{x\sqrt{x\sqrt{\cdots}}}} \implies S^2 = xS \implies S = x.$

2) $\sqrt{x\sqrt{x\sqrt{x\sqrt{\cdots}}}} = x^{\frac{1}{2}}x^{\frac{1}{4}}x^{\frac{1}{8}}\cdots = x^{\frac{1}{2}+\frac{1}{4}+\frac{1}{8}+\cdots} = x^1 = x$

7) Find the value of $\sqrt{5 + \sqrt{5^2 + \sqrt{5^4 + \sqrt{5^8 + \dots}}}}$

$$\sqrt{5 + \sqrt{5^2 + \sqrt{5^4 + \sqrt{5^8 + \dots}}}}$$

$$= \sqrt{5} \times \sqrt{1 + \sqrt{1 + \sqrt{1 + \sqrt{1 + \dots}}}}$$

$$= \sqrt{5} \times \frac{\sqrt{5}+1}{2}$$

$$= \boxed{\frac{5 + \sqrt{5}}{2}}$$

8) Compute

$$\sqrt{\frac{2}{2^2} + \sqrt{\frac{2}{2^4} + \sqrt{\frac{2}{2^8} + \cdots}}}$$

$$\sqrt{\frac{2}{2^2} + \sqrt{\frac{2}{2^4} + \sqrt{\frac{2}{2^8} + \cdots}}} = \frac{1}{2}\sqrt{2 + \sqrt{2 + \sqrt{2 + \cdots}}} = \frac{1}{2} \times 2 = 1$$

9) Without using a calculator, explain that

$$\sqrt{20 + \sqrt{20 + \sqrt{20}}} - \sqrt{20 - \sqrt{20 - \sqrt{20}}} \approx 1$$

This approximation holds because:

$$\sqrt{20 + \sqrt{20 + \sqrt{20 + \sqrt{\cdots}}}} = 5$$

$$\sqrt{20 - \sqrt{20 - \sqrt{20 - \sqrt{\cdots}}}} = 4$$

Therefore

$$\sqrt{20 + \sqrt{20 + \sqrt{20}}} - \sqrt{20 - \sqrt{20 - \sqrt{20}}}$$

$$\approx \sqrt{20 + \sqrt{20 + \sqrt{20 + \sqrt{\cdots}}}} - \sqrt{20 - \sqrt{20 - \sqrt{20 - \sqrt{\cdots}}}}$$

$$= 5 - 4$$

$$= 1$$

10) Srinivasa Ramanujan, an Indian mathematician, claimed that the following relationship held:

$$x + n = \sqrt{n^2 + x\sqrt{n^2 + (x + n)\sqrt{n^2 + (x + 2n)\sqrt{\cdots}}}} \qquad (15.4)$$

Do you agree with him? Explain.

Solution

This claim holds because

$$x + n$$
$$= \sqrt{(n+x)^2}$$
$$= \sqrt{n^2 + x(x+n+n)}$$
$$= \sqrt{n^2 + x\sqrt{(n+(x+n))^2}}$$
$$= \sqrt{n^2 + x\sqrt{n^2 + (x+n)(x+n+2n)}}$$
$$= \sqrt{n^2 + x\sqrt{n^2 + (x+n)\sqrt{(n+(x+2n))^2}}}$$
$$= \cdots$$

11) Evaluate

$$\sqrt{1 + 2\sqrt{1 + 3\sqrt{1 + 4\sqrt{1 + \cdots}}}}$$

Letting $n = 1$ and $x = 2$ in *Equation 15.4* in the previous practice leads to the answer $\boxed{3}$.

Lecture 6

1) Simplify $\sqrt{10 + 4\sqrt{3 - 2\sqrt{2}}}$.

$$\sqrt{10 + 4\sqrt{3 - 2\sqrt{2}}}$$
$$=\sqrt{10 + 4(\sqrt{2} - 1)}$$
$$=\sqrt{6 - 4\sqrt{2}}$$
$$=\sqrt{2}\sqrt{3 + 2\sqrt{2}}$$
$$=\sqrt{2}(\sqrt{2} + 1)$$
$$=\boxed{2 + \sqrt{2}}$$

2) Evaluate $\sqrt{7 + 4\sqrt{3}} + \sqrt{7 - 4\sqrt{3}}$.

$$\sqrt{7 + 4\sqrt{3}} + \sqrt{7 - 4\sqrt{3}} = (2 + \sqrt{3}) + (2 - \sqrt{3}) = \boxed{4}$$

3) Evaluate
$$\left(\sqrt{3 - 2\sqrt{2}} + \sqrt{3 + 2\sqrt{2}}\right)^2$$

$$\left(\sqrt{3 - 2\sqrt{2}} + \sqrt{3 + 2\sqrt{2}}\right)^2 = \left((\sqrt{2}-1)+(\sqrt{2}+1)\right)^2 = (2\sqrt{2})^2 = \boxed{8}$$

4) Simplify $\sqrt{\sqrt[3]{9} + 6\sqrt[3]{3} + 9}$.

$$\sqrt{\sqrt[3]{9} + 6\sqrt[3]{3} + 9} = \sqrt{(\sqrt[3]{3})^2 + 2 \times 3 \times \sqrt[3]{3} + 3^2} = \boxed{3 + \sqrt[3]{3}}$$

> ⓘ *Tip: when there are three terms inside a radical, such as $\sqrt{A + B + C}$ where two or three of $A, B,$ and C are radicals themselves, it simply means that at least one term in the simplified form is not a square root. For example, in this problem, $\sqrt[3]{3}$ still contains radical after being squared.*

5) Simplify $\sqrt{4 + \sqrt[3]{81} + 4\sqrt[3]{9}}$.

$$\sqrt{4 + \sqrt[3]{81} + 4\sqrt[3]{9}} = \sqrt{2^2 + (\sqrt[3]{9})^2 + 2 \times 2 \times \sqrt[3]{9}} = \boxed{2 + \sqrt[3]{9}}$$

6) Simplify $\sqrt{6 + \sqrt[3]{81} + \sqrt[3]{9}}$.

$$\sqrt{6 + \sqrt[3]{81} + \sqrt[3]{9}} = \sqrt{2 \times \sqrt[3]{9} \times \sqrt[3]{3} + (\sqrt[3]{9})^2 + (\sqrt[3]{3})^2} = \boxed{\sqrt[3]{3} + \sqrt[3]{9}}$$

7) Let $\sqrt{1 + \sqrt{21 + 12\sqrt{3}}} = \sqrt{a} + \sqrt{b}$. Find $a + b$.

$$\sqrt{1 + \sqrt{21 + 12\sqrt{3}}}$$
$$= \sqrt{1 + \sqrt{21 + 2\sqrt{108}}} =$$

$$=\sqrt{1+(3+2\sqrt{3})}$$
$$=\sqrt{4+2\sqrt{3}}$$
$$=1+\sqrt{3}$$

Therefore the answer is $1+3=\boxed{4}$.

8) Solve the equation $x^2+6x-4\sqrt{5}=0$.

By factorization, we find $x^2+6x-4\sqrt{5}=(x+1-\sqrt{5})(x+5+\sqrt{5})$.
Therefore the two roots are $\boxed{-1+\sqrt{5}}$ and $\boxed{-5-\sqrt{5}}$.
Alternatively, by using the quadratic formula, we find

$$x_{1,2}=\frac{-6\pm\sqrt{36+16\sqrt{5}}}{2}=-3\pm\sqrt{9+4\sqrt{5}}$$

Note that $\sqrt{9+4\sqrt{5}}=\sqrt{9+2\sqrt{20}}=2+\sqrt{5}$. Simplifying the above result will lead to the same answer.

9) Simplify $\sqrt{5\sqrt{3}+6\sqrt{2}}$.

Note that $5=3+2$.

$$\therefore \quad \sqrt{5\sqrt{3}+6\sqrt{2}}$$
$$=\sqrt{\sqrt{3}\times(3+2\sqrt{2}\sqrt{3}+2)}$$
$$=\sqrt[4]{3}\times(\sqrt{3}+\sqrt{2})$$
$$=\sqrt[4]{27}+\sqrt[4]{12}$$

10) Simplify $(\sqrt[3]{3} + \sqrt[3]{2})(\sqrt[3]{9} - \sqrt[3]{6} + \sqrt[3]{4})$

By the polynomial identity: $(a + b)(a^2 - ab + b^2) = a^3 + b^3$, we have

$$(\sqrt[3]{3} + \sqrt[3]{2})(\sqrt[3]{9} - \sqrt[3]{6} + \sqrt[3]{4}) = (\sqrt[3]{3})^3 + (\sqrt[3]{2})^3 = \boxed{5}$$

ⓘ   *Tip: polynomial identity can be a great tool.*

11) Compute $\sqrt{1 + 1995\sqrt{4 + 1995 \times 1999}}$.

Note that by difference of squares, we have

$$(1997 - 2)(1997 + 2) = 1995 \times 1999 = 1997^2 - 4$$

Therefore

$$\sqrt{1 + 1995\sqrt{4 + 1995 \times 1999}}$$
$$= \sqrt{1 + 1995\sqrt{4 + (1997^2 - 4)}}$$
$$= \sqrt{1 + 1995 \times 1997}$$
$$= \sqrt{1 + (1996^2 - 1)}$$
$$= \boxed{1996}$$

12) Simplify $\sqrt{12 + 2\sqrt{6} + 2\sqrt{14} + 2\sqrt{21}}$.

Note that $(a + b + c)^2 = a^2 + b^2 + c^2 + 2ab + 2bc + 2ca$ and $6 = 2 \times 3, 14 = 2 \times 7, 21 = 3 \times 7$, and $2 + 3 + 7 = 12$. It is then easy to verify:

$$\sqrt{12 + 2\sqrt{6} + 2\sqrt{14} + 2\sqrt{21}} = \sqrt{2} + \sqrt{3} + \sqrt{7}$$

Lecture 7

1) Compute the value of the following infinite expression:

$$\frac{1}{1 \times 2} + \frac{1}{2 \times 3} + \cdots$$

Let $S_n = \frac{1}{1\times2} + \frac{1}{2\times3} + \cdots + \frac{1}{n\times(n+1)} = 1 - \frac{1}{n+1}$.

Then the desired answer is simply the value of S_n when n approaches infinity which is $\boxed{1}$.

2) Suppose function $f(x) = \frac{1+x}{1-x}$. Evaluate

$$f\left(\frac{1}{2}\right) \cdot f\left(\frac{1}{4}\right) \cdot f\left(\frac{1}{6}\right) \cdots f\left(\frac{1}{2014}\right)$$

(Ref 2013 China)

$$f\left(\frac{1}{2}\right) \cdot f\left(\frac{1}{4}\right) \cdot f\left(\frac{1}{6}\right) \cdots f\left(\frac{1}{2014}\right)$$

$$= \frac{3}{1} \times \frac{5}{3} \times \frac{7}{5} \times \cdots \times \frac{2015}{2013}$$

$$= \boxed{2015}$$

3) Compute

$$\frac{1}{1 \times 3} + \frac{1}{3 \times 5} + \cdots + \frac{1}{2015 \times 2017}$$

$$\frac{1}{1 \times 3} + \frac{1}{3 \times 5} + \cdots + \frac{1}{2015 \times 2017}$$

$$= \frac{1}{2} \times \left(\frac{1}{1} - \frac{1}{3}\right) + \frac{1}{2} \times \left(\frac{1}{3} - \frac{1}{5}\right) + \cdots + \frac{1}{2} \times \left(\frac{1}{2015} - \frac{1}{2017}\right)$$

$$= \frac{1}{2} \times \left(\frac{1}{1} - \frac{1}{2017}\right)$$

$$= \frac{1008}{2017}$$

ⓘ   *Tip: As long as the numbers in the denominator forms an arithmetic sequence, the technique discussed in this lecture can be used.*

4) Compute

$$\sum_{k=1}^{\infty} \frac{1}{k^2 + k}$$

$$\sum_{k=1}^{\infty} \frac{1}{k^2 + k}$$

$$= \sum_{k=1}^{\infty} \left(\frac{1}{k} - \frac{1}{k+1}\right)$$

$$= \left(1 - \frac{1}{2}\right) + \left(\frac{1}{2} - \frac{1}{3}\right) + \left(\frac{1}{3} - \frac{1}{4}\right) + \cdots$$

$$= 1$$

5) Evaluate the infinite sum $\displaystyle\sum_{n=1}^{\infty} \frac{n}{n^4 + 4}$.

$$\sum_{n=1}^{\infty} \frac{n}{n^4 + 4}$$

$$= \sum_{n=1}^{\infty} \frac{n}{(n^2 + 2n + 2)(n^2 - 2n + 2)}$$

$$= \frac{1}{4} \times \sum_{n=1}^{\infty} \left(\frac{1}{n^2 - 2n + 2} - \frac{1}{n^2 + 2n + 2} \right)$$

$$= \frac{1}{4} \times \sum_{n=1}^{\infty} \left(\frac{1}{(n-1)^2 + 1} - \frac{1}{(n+1)^2 + 1} \right)$$

$$= \frac{1}{4} \times \left(\frac{1}{0^2 + 1} + \frac{1}{1^2 + 1} \right)$$

$$= \boxed{\frac{3}{8}}$$

6) Let S_n be the sum of first n terms in sequence $\{a_n\}$ where

$$a_n = \sqrt{1 + \frac{1}{n^2} + \frac{1}{(n+1)^2}}$$

Find $\lfloor S_n \rfloor$ where the floor function $\lfloor x \rfloor$ returns the largest integer not exceeding x.

Firstly, we note that

$$1 + \frac{1}{n^2} + \frac{1}{(n+1)^2} = \left(1 + \frac{1}{n} - \frac{1}{n+1} \right)^2$$

This identity holds because

$$\left(1 + \frac{1}{n} - \frac{1}{n+1} \right)^2 = 1 + \frac{1}{n^2} + \frac{1}{(n+1)^2} + 2 \times \left(\frac{1}{n} - \frac{1}{n+1} - \frac{1}{n(n+1)} \right)$$

and

$$\frac{1}{n(n+1)} = \frac{1}{n} - \frac{1}{n+1}$$

Therefore

$$a_n = 1 + \frac{1}{n} - \frac{1}{n+1}$$

$$S_n = \left(1 + 1 - \frac{1}{2}\right) + \left(1 + \frac{1}{2} - \frac{1}{3}\right) + \cdots + \left(1 + \frac{1}{n} - \frac{1}{n+1}\right)$$
$$= n + \left(1 - \frac{1}{n+1}\right)$$

$$\therefore 0 < 1 - \frac{1}{n+1} < 1 \implies \lfloor S_n \rfloor = \boxed{n}$$

7) Let $\{a_n\}$ be an increasing geometric sequence satisfying $a_1 + a_2 = 6$ and $a_3 + a_4 = 24$. Let $\{b_n\}$ be another sequence satisfying $b_n = \frac{a_n}{(a_n-1)^2}$. If T_n is the sum of first n terms in $\{b_n\}$, show that for any positive integer n, the relation $T_n < 3$ always holds.

Let r be the common ratio of $\{a_n\}$. Then

$$\frac{a_3 + a_4}{a_1 + a_2} = \frac{a_1 r^2 + a_2 r^2}{a_1 + a_2} \implies r^2 = \frac{24}{6} = 4 \implies r = 2 \quad (\because r > 1)$$

It follows that

$$a_1 + a_2 = a_1(1 + r) = 3a_1 \implies 3a_1 = 6 \implies a_1 = 2$$

$$\therefore \quad a_n = a_1 r^{n-1} = 2 \times 2^{n-1} = 2^n$$

Hence,

$$b_n = \frac{a_n}{(a_n - 1)^2}$$

Solution

$$= \frac{2^n}{(2^n - 1)^2}$$

$$< \frac{2^n}{(2^n - 1)(2^n - 2)}$$

$$= \frac{2^{n-1}}{(2^n - 1)(2^{n-1} - 1)}$$

It is easy to verify that

$$\frac{2^{n-1}}{(2^{n-1} - 1)(2^n - 1)} = \frac{1}{2^{n-1} - 1} - \frac{1}{2^n - 1}$$

Therefore

$$T_n = b_1 + b_2 + \cdots + b_n$$

$$< 2 + \left(1 - \frac{1}{3}\right) + \left(\frac{1}{3} - \frac{1}{7}\right) + \cdots + \left(\frac{1}{2^{n-1} - 1} - \frac{1}{2^n - 1}\right)$$

$$= 3 - \frac{1}{2^n - 1} < 3$$

---

8) For $n \geq 1$, let d_n denote the length of the line segment connecting the two points where the line $y = x + n + 1$ intersects the parabola $8x^2 = y - \frac{1}{32}$. Compute the sum

$$\sum_{n=1}^{1000} \frac{1}{n \cdot d_n^2}$$

Canceling y in $y = x + n + 1$ and $8x^2 = y - \frac{1}{32}$ leads to

$$8x^2 - x - \left(n + \frac{31}{32}\right) = 0$$

Therefore

$$x_{1,2} = \frac{1}{16} \pm \frac{1}{4} \times \sqrt{2n + 2} \implies |x_1 - x_2| = \frac{1}{2} \times \sqrt{2n + 2}$$

Clearly, the slop of $y = x + n + 1$ is 1. Hence we find

$$d_n = \left(\frac{1}{2} \times \sqrt{2n+2}\right) \times \sqrt{2} = \sqrt{n+1}$$

It follows that

$$\frac{1}{n \cdot d_n^2} = \frac{1}{n(n+1)} \implies \sum_{n=1}^{1000} \frac{1}{n \cdot d_n^2} = \frac{1}{1} - \frac{1}{1001} = \boxed{\frac{1000}{1001}}$$

9) Compute the value of

$$\sum_{n=1}^{\infty} \frac{2n+1}{n^2(n+1)^2}$$

Suppose

$$\frac{2n+1}{n^2(n+1)^2} = \frac{A}{n^2} + \frac{B}{(n+1)^2}$$

Then we have

$$2n + 1 = A(n+1)^2 + Bn^2$$

Let $n = 0 \implies A = 1$ and Let $n = -1 \implies B = -1$.

$$\therefore \quad \frac{2n+1}{n^2(n+1)^2} = \frac{1}{n^2} - \frac{1}{(n+1)^2}$$

Hence,

$$\sum_{n=1}^{\infty} \frac{2n+1}{n^2(n+1)^2} = \left(\frac{1}{1^2} - \frac{1}{2^2}\right) + \left(\frac{1}{2^2} - \frac{1}{3^2}\right) + \cdots = 1$$

10) Find the length of the leading non-repeating block in the decimal expansion of $\frac{2017}{3 \times 5^{2016}}$. For example the length of the leading non-repeating block of $\frac{1}{6} = 0.1\overline{6}$ is 1.

Notice that $\frac{2017}{3 \times 5^{2016}}$ can be decompose to

$$\frac{2017}{3 \times 5^{2016}} = \frac{A}{3} + \frac{B}{5^{2016}}$$

where A and B are two constants.

It is apparent that $\frac{A}{3}$ is a recurring decimal. But $\frac{B}{5^{2016}}$ is a finite decimal which has 2016 digits to the right of the decimal point. Therefore their sum must have a leading 2016 non-repeating block.

Lecture 8

1) Simplify

$$\frac{1}{1 \times 2 \times 3} + \frac{1}{2 \times 3 \times 4} + \cdots + \frac{1}{n \times (n+1) \times (n+2)}$$

This is a generalized result of *Example 8.2.1* on *page 32*. The answer is:

$$\frac{1}{2} \times \left(\frac{1}{1 \times 2} - \frac{1}{(n+1)(n+2)} \right) = \frac{n(n+3)}{4(n+1)(n+2)}$$

2) Compute the value of the following infinite expression:

$$\frac{1}{1 \times 2 \times 3} + \frac{1}{2 \times 3 \times 4} + \cdots$$

By the previous practice, the desired result is:

$$\lim_{n \to \infty} \left[\frac{1}{2} \times \left(\frac{1}{1 \times 2} - \frac{1}{(n+1)(n+2)} \right) \right] = \boxed{\frac{1}{4}}$$

3) Use the *special value method* to determine coefficient A and B in the identities below:

$$\frac{1}{k \times (k+1)} = \frac{A}{k} + \frac{B}{k+1}$$

$$\frac{1}{k \times (k+2)} = \frac{A}{k} + \frac{B}{k+2}$$

1) First, let's transform the given expression

$$\frac{1}{k \times (k+1)} = \frac{A}{k} + \frac{B}{k+1} \implies A(k+1) + Bk = 1$$

Then, lettings $k = 0 \implies A = 1$, letting $k = -1 \implies B = -1$

Hence

$$\frac{1}{k \times (k+1)} = \frac{1}{k} - \frac{1}{k+1}$$

2) This can be solved in a similar way

$$\frac{1}{k \times (k+2)} = \frac{A}{k} + \frac{B}{k+2} \implies A(k+2) + Bk = 1$$

Then, lettings $k = 0 \implies A = \frac{1}{2}$, letting $k = -2 \implies B = -\frac{1}{2}$

Hence

$$\frac{1}{k \times (k+2)} = \frac{1}{2} \times \left(\frac{1}{k} - \frac{1}{k+2} \right)$$

4) Evaluate the value of

$$\sum_{n=1}^{\infty} \frac{6}{(2n-1)(2n+1)}$$

Suppose

$$\frac{6}{(2n-1)(2n+1)} = \frac{A}{2n-1} + \frac{B}{2n+1} \qquad (15.5)$$

where A and B are to be determined constants. *(15.5)* can be re-written as

$$(2n+1)A + (2n-1)B = 6$$

$$\text{Let} \quad 2n+1 = 0 \implies n = -\frac{1}{2} \implies B = -3$$

$$\text{Let} \quad 2n-1 = 0 \implies n = \frac{1}{2} \implies A = 3$$

$$\therefore \quad \frac{6}{(2n-1)(2n+1)} = \frac{3}{2n-1} - \frac{3}{2n+1}$$

$$\therefore \quad \sum_{n=1}^{\infty} \frac{6}{(2n-1)(2n+1)}$$

$$= \sum_{n=1}^{\infty} \left(\frac{3}{2n-1} - \frac{3}{2n+1} \right)$$

$$= \left(\frac{3}{1} - \frac{3}{3} \right) + \left(\frac{3}{3} - \frac{3}{5} \right) + \cdots$$

$$= 3$$

---

5) Compute

$$\sum_{n=1}^{\infty} \frac{2}{n^2 + 4n + 3}$$

$$\therefore \quad \frac{2}{n^2 + 4n + 3} = \frac{2}{(n+1)(n+3)} = \frac{1}{n+1} - \frac{1}{n+3}$$

$$\therefore \quad \sum_{n=1}^{\infty} \frac{2}{n^2 + 4n + 3} = \frac{1}{1+1} + \frac{1}{2+1} = \frac{5}{6}$$

---

6) Find the remainder when $x^{81} + x^{49} + x^{25} + x^9 + x$ is divided by $x^3 - x$.

Because the degree of $(x^3 - x)$ is 3, the desired reminder is at most a quadratic polynomial. Let the quotient be $q(x)$, and the remainder be $r(x) = ax^2 + bx + c$. Then

$$x^{81} + x^{49} + x^{25} + x^9 + x = q(x)(x^3 - x) + r(x)$$

Solution

Plugging in the values $x = -1, 0, 1$ which are the three roots of $(x^3 - x)$ yields $r(-1) = -5, r(0) = 0, r(1) = 5$, or

$$\begin{cases} a - b + c & = & -5 \\ c & = & 0 \\ a + b + c & = & 5 \end{cases}$$

From here we get $a = c = 0, b = 5$, hence the remainder is

$$r(x) = ax^2 + bx + c = \boxed{5x}$$

Lecture 9

1) Show that

$$C_n^0 + 2C_n^1 + 4C_n^2 + \cdots + 2^n C_n^n = 3^n$$

Letting $x = 2$ in $(1 + x)^n$ will lead to the desired result.

2) Compute
$$C_n^1 + 2C_n^2 + \cdots + nC_n^n$$

This problem can be solved using the reversing order method. Let the sum be S then:

$$S = \quad 0 \times C_n^0 \quad + \qquad\qquad 1 \times C_n^1 \quad + \quad \cdots \quad + \quad n \times C_n^n$$
$$S = \quad n \times C_n^n \quad + \quad (n-1) \times C_n^{n-1} \quad + \quad \cdots \quad + \quad 0 \times C_n^n$$

Note that $C_n^0 = C_n^n, C_n^1 = C_n^{n-1}$ and so on, adding these two equations yields:

$$2S = n \times (C_n^0 + C_n^1 + \cdots + C_n^n) = n \times 2^n \implies S = \boxed{n \times 2^{n-1}}$$

3) If the sum of all coefficients in the expanded form of $(3x + 1)^n$ is 256, find the coefficient of x^2.

Solution

The sum of all coefficients in the expanded form of $(3x+1)^n$ can be computed by setting x as 1. Therefore,

$$(3 \times 1 + 1)^n = 256 \implies n = 4$$

By the binomial theorem, the term containing x^2 is

$$C_4^2(3x)^2 = 6 \times 3^2 x^2 = 54x^2$$

Hence, we conclude the coefficient of x^2 is $\boxed{54}$.

---

4) Find the coefficient of x^{17} in the expansion of $(1 + x^5 + x^7)^{20}$.

(Ref 2001 HK Team Selection)

The only way to get x^{17} is to multiply two x^5 and one x^7. Hence the answer is

$$C_{20}^2 C_{18}^1 = 3420$$

This is equivalent to, firstly, selecting 2 of the $(1 + x^5 + x^7)$ terms out of 20 to contribute x^5 each, and then selecting 1 of the remaining 18 terms to contribute an x^7 element[3].

---

5) **(Vandermonde Identity)** Given positive integers m, n, and r, where $r \le m$ and $r \le n$, prove the following relation always holds:

$$C_m^0 C_n^r + C_m^1 C_n^{r-1} + \cdots + C_m^r C_n^0 = C_{m+n}^r$$

---

[3]This is a typical counting problem. Solving competition counting problems is discussed in the book *Counting* by the same author.

Consider the coefficient of x^r in the expanded form of the following identity.

$$(1+x)^{m+n} = (1+x)^m(1+x)^n$$

On the left side, the coefficient is clearly C_{m+n}^r.

The right side can be expanded into

$$(1+x)^m(1+x)^n = (C_m^0 + C_m^1 x + \cdots + C_m^m x^m)(C_n^0 + C_n^1 x + \cdots + C_n^n x^n)$$

Therefore the coefficient of x^r will be

$$C_m^0 C_n^r + C_m^1 C_n^{r-1} + \cdots + C_m^r C_n^0$$

These two values must equal. Hence we conclude

$$C_{m+n}^r = C_m^0 C_n^r + C_m^1 C_n^{r-1} + \cdots + C_m^r C_n^0 = C_{m+n}^r$$

6) Simplify $\displaystyle\sum_{k=0}^{n} \left(C_n^k\right)^2 = \left(C_n^0\right)^2 + \left(C_n^1\right)^2 + \cdots + \left(C_n^n\right)^2$.

$$\sum_{k=0}^{n} \left(C_n^k\right)^2 = \left(C_n^0\right)^2 + \left(C_n^1\right)^2 + \cdots + \left(C_n^n\right)^2$$
$$= C_n^0 C_n^n + C_n^1 C_n^{n-1} + \cdots + C_n^n C_n^0$$
$$= C_{2n}^n \qquad \text{by the Vandermonde's identity}$$

7) Show that

$$\sum_{i=1}^{n} C_n^i C_n^{i-1} = C_{2n}^{n-1}$$

$$\sum_{i=1}^{n} C_n^i C_n^{i-1} = \sum_{i=0}^{n-1} C_n^{i+1} C_n^i = \sum_{i=0}^{n-1} C_n^{n-1-i} C_n^i = C_{2n}^{n-1}$$

The last step uses the Vandermonde identity.

8) Use the binomial theorem to prove that $3^{4n+2} + 5^{2n+1}$ is divisible by 14 for any given positive integer n.

$$3^{4n+2} + 5^{2n+1}$$
$$= 9^{2n+1} + 5^{2n+1}$$
$$= [(9+5) - 5)]^{2n+1} + 5^{2n+1}$$
$$= \left(C_{2n+1}^0 14^{2n+1} - C_{2n+1}^1 14^{2n} \cdot 5 + \cdots + C_{2n+1}^{2n} \cdot 14 \cdot 5^{2n} \right.$$
$$\left. - C_{2n+1}^{2n+1} \cdot 5^{2n+1} \right) + 5^{2n+1}$$
$$= C_{2n+1}^0 14^{2n+1} - C_{2n+1}^1 14^{2n} \cdot 5 + \cdots + C_{2n+1}^{2n} \cdot 14 \cdot 5^{2n}$$

It is apparent that every term is a multiple of 14. Therefore the claim holds.

9) Let the integer and decimal part of $(5\sqrt{2} + 7)^{2n+1}$ be I and D respectively. Show that $(I + D) \cdot D$ is a constant.

By the binomial theorem, it is apparent that

$$(5\sqrt{2} + 7)^{2n+1} - (5\sqrt{2} - 7)^{2n+1}$$

is an integer.

Meanwhile

$$0 < 5\sqrt{2} - 7 < 1 \implies 0 < (5\sqrt{2} - 7)^{2n+1} < 1$$

This means

$$D = (5\sqrt{2} - 7)^{2n+1} \quad \text{and} \quad I + D = (5\sqrt{2} + 7)^{2n+1}$$

Therefore

$$(I + D) \cdot D = (5\sqrt{2} + 7)^{2n+1}(5\sqrt{2} - 7)^{2n+1} = 1$$

which is a constant.

10) Let a, b be two positive real numbers, and n be a positive integer greater than 2. Show that

$$\frac{a^n + a^{n-1}b + \cdots + ab^{-1} + b^n}{n+1} \geq \left(\frac{a+b}{2}\right)^n$$

The claim obviously holds when $a = b$.

When $a \neq b$, the original claim is equivalent to

$$\frac{a^{n+1} - b^{n+1}}{(n+1)(a-b)} \geq \left(\frac{a+b}{2}\right)^n$$

Let $x = \frac{a+b}{2}$ and $y = \frac{a-b}{2}$. Then $a = x + y, b = x - y$.

$$\frac{a^{n+1} - b^{n+1}}{(n+1)(a-b)}$$

$$= \frac{1}{2y(n+1)}\left[(x+y)^{n+1} - (x-y)^{n+1}\right]$$

$$= \frac{1}{n+1}\left(C_{n+1}^1 x^n + C_{n+1}^3 x^{n+2} + \cdots\right)$$

$$\geq \frac{1}{n+1}C_{n+1}^1 x^n \qquad \because x > 0$$

$$= \left(\frac{a+b}{2}\right)^n$$

Lecture 10

1) Evaluate

$$C_n^0 - C_n^1 + C_n^2 - C_n^3 + \cdots \qquad (15.6)$$

The answer is 0 because by *Example 10.2.1* on *page 41*:

$$C_n^0 + C_n^2 + C_n^4 + \cdots = C_n^1 + C_n^3 + C_n^5 + \cdots = 2^{n-1}$$

2) Show that

$$C_n^0 - C_n^2 + C_n^4 - C_n^6 + \cdots = 2^{\frac{n}{2}} \cos \frac{n\pi}{4}$$

$$C_n^1 - C_n^3 + C_n^5 - C_n^7 + \cdots = 2^{\frac{n}{2}} \sin \frac{n\pi}{4}$$

Let $x = i$ in the expanded form of $(1+x)^n$:

$$(1+x)^n = (C_n^0 - C_n^2 + C_n^4 - C_n^6 + \cdots) + i(C_n^1 - C_n^3 + C_n^5 - C_n^7 + \cdots)$$

On the other hand,

$$(1+i)^n = \left(\sqrt{2}\left(\cos \frac{\pi}{4} + i \sin \frac{\pi}{4}\right)\right)^n = 2^{\frac{n}{2}}\left(\cos \frac{n\pi}{4} + i \sin \frac{n\pi}{4}\right)$$

Comparing the real and imaginary parts of the above two equations leads to the desired claims.

3) Simplify the following expressions:

(i) $C_{2016}^0 + C_{2016}^3 + C_{2016}^6 + \cdots + C_{2016}^{2016}$

(ii) $C_{2016}^1 + C_{2016}^4 + C_{2016}^7 + \cdots + C_{2016}^{2014}$

(iii) $C_{2016}^2 + C_{2016}^5 + C_{2016}^8 + \cdots + C_{2016}^{2016}$

Let ω be one complex root of the equation $x^3 = 1$. Then, it must hold that $\omega^3 = 1$ and $1 + \omega + \omega^2 = 0$.

Setting $x = 1, \omega$, and ω^2 in

$$(1 + x)^n = C_n^0 + C_n^1 x + C_n^2 x^2 + \cdots + C_n^n x^n$$

, respectively, leads:

$$
\begin{array}{lllll}
(1+1)^n & = & C_n^0 + & C_n^1 + & C_n^2 + \cdots & (1) \\
(1+\omega)^n & = & C_n^0 + & \omega C_n^1 + & \omega^2 C_n^2 + \cdots & (2) \\
(1+\omega^2)^n & = & C_n^0 + & \omega^2 C_n^1 + & \omega^4 C_n^2 + \cdots & (3)
\end{array}
$$

Applying $\omega^3 = 1$ and $1 + \omega + \omega^2 = 0$, $(1), (2)$, and (3) can be rewritten as:

$$
\begin{array}{llllll}
2^n & = & C_n^0 + & C_n^1 + & C_n^2 + & C_n^3 + \cdots & (4) \\
(-\omega^2)^n & = & C_n^0 + & \omega C_n^1 + & \omega^2 C_n^2 + & C_n^3 + \cdots & (5) \\
(-\omega)^n & = & C_n^0 + & \omega^2 C_n^1 + & \omega C_n^2 + & C_n^3 + \cdots & (6)
\end{array}
$$

$(4) + (5) + (6)$ yields:

$$
\begin{aligned}
& 2^n + (-1)^n \omega^{2n} + (-1)^n \omega^n \\
= & 3C_n^0 + (1 + \omega + \omega^2)C_n^1 + (1 + \omega^2 + \omega)C_n^2 + 3C_n^3 + \cdots \\
= & 3C_n^0 + 3C_n^3 + \cdots
\end{aligned}
$$

$$\therefore \quad C_n^0 + C_n^3 + \cdots = \frac{1}{3}\left(2^n + (-1)^n \omega^{2n} + (-1)^n \omega^n\right) \quad (7)$$

$(4) + \omega \cdot (5) + \omega^2 \cdot (6)$ yields:

$$
\begin{aligned}
& 2^n + (-1)^n \omega^{2n+1} + (-1)^n \omega^{n+2} \\
= & (1 + \omega + \omega^2)C_n^0 + (1 + \omega^2 + \omega)C_n^1 + 3C_n^2 + \cdots \\
= & 3C_n^2 + 3C_n^5 + \cdots
\end{aligned}
$$

$$\therefore \quad C_n^2 + C_n^5 + \cdots = \frac{1}{3}\left(2^n + (-1)^n \omega^{2n+1} + (-1)^n \omega^{n+2}\right) \quad (8)$$

$(4) + \omega^2 \cdot (5) + \omega \cdot (6)$ yields:

$$2^n + (-1)^n \omega^{2n+2} + (-1)^n \omega^{n+1}$$
$$= (1 + \omega^2 + \omega)C_n^0 + 3C_n^1 + (1 + \omega + \omega^2)C_n^2 + \cdots$$
$$= 3C_n^1 + 3C_n^4 + \cdots$$

$$\therefore \quad C_n^1 + C_n^4 + \cdots = \frac{1}{3}\left(2^n + (-1)^n \omega^{2n+2} + (-1)^n \omega^{n+1}\right) \qquad (9)$$

Setting $n = 2016$ in $(7), (8)$, and (9) yields:

$$C_{2016}^0 + C_{2016}^3 + \cdots + C_{2016}^{2016} = \tfrac{1}{3} \times \left(2^{2016} + 2\right)$$

$$C_{2016}^1 + C_{2016}^4 + \cdots + C_{2016}^{2014} = \tfrac{1}{3} \times \left(2^{2016} - 1\right)$$

$$C_{2016}^2 + C_{2016}^5 + \cdots + C_{2016}^{2015} = \tfrac{1}{3} \times \left(2^{2016} - 1\right)$$

4) If $a_n = C_{2003}^{3n-1}$, find the value of $\displaystyle\sum_{n=1}^{668} a_n$.

(Ref 2003 China)

This is a special case of the previous problem. Therefore

$$\sum_{n=1}^{668} C_{2003}^{3n-1} = \frac{1}{3} \times \left(2^{2003} + (-1)^{2003}\omega^{4007} + (-1)^{2003}\omega^{2005}\right)$$

$$= \frac{1}{3} \times \left(2^{2003} - \omega^2 - \omega\right)$$

$$= \frac{1}{3} \times \left(2^{2003} + 1\right)$$

5) Simplify

$$C_n^0 - \frac{1}{2}C_n^1 + C_n^2 - \frac{1}{2}C_n^3 + \cdots$$

We know that

$$C_n^0 + C_n^2 + \cdots = C_n^1 + C_n^3 + \cdots = 2^{n-1}$$

Therefore

$$C_n^0 - \frac{1}{2}C_n^1 + C_n^2 - \frac{1}{2}C_n^3 + \cdots$$

$$= (C_n^0 + C_n^2 + \cdots) - \frac{1}{2} \times (C_n^1 + C_n^3 + \cdots)$$

$$= 2^{n-1} - \frac{1}{2} \times 2^{n-1}$$

$$= \boxed{2^{n-2}}$$

Lecture 11

1) When preparing goodie bags for his birthday party, Joe randomly puts between 20 and 30 M&M chocolates in each bag. Suppose M&M has 5 different colors available, how many different goodie bags are possible?

If there are 20 M&Ms in a bag, there will be totally

$$C^4_{20+5-1} = C^4_{24}$$

different types of bags. This is equivalent to finding non-negative integer solutions to the following equation

$$x_1 + x_2 + x_3 + x_4 + x_5 = 20$$

Solving such problems is discussed in the book *Counting* by the same author.

Therefore the desired answer is

$$C^4_{24} + C^4_{25} + \cdots + C^4_{34}$$

This can be computed by using the Hockey Stick identity:

$$
\begin{aligned}
& C^4_{24} + C^4_{25} + \cdots + C^4_{34} \\
=& (C^4_4 + C^4_5 + \cdots + C^4_{34}) - (C^4_4 + C^4_5 + \cdots + C^4_{23}) \\
=& C^5_{35} - C^5_{24}
\end{aligned}
$$

---

2) Let $n > k$ be two positive integers. Simplify

$$C^k_n + 2C^k_{n-1} + 3C^k_{n-2} + \cdots (n-k+1)C^k_k$$

$$C_n^k + 2C_{n-1}^k + 3C_{n-2}^k + \cdots + (n-k+1)C_k^k$$
$$= \left(C_n^k + C_{n-1}^k + C_{n-2}^k + \cdots + C_k^k \right) +$$
$$\left(C_{n-1}^k + C_{n-2}^k + \cdots + C_k^k \right) +$$
$$\left(C_{n-2}^k + \cdots + C_k^k \right) +$$
$$\cdots$$
$$C_k^k$$
$$= C_{n+1}^{k+1} + C_n^{k+1} + C_{n-1}^{k+1} + \cdots + C_{k+1}^{k+1} \qquad \because (C_k^k = C_{k+1}^{k+1})$$
$$= C_{n+2}^{k+2}$$

---

3) Using the Hockey Stick identity to compute $1 + 2 + 3 + \cdots + n$.

$$1 + 2 + 3 + \cdots + n$$
$$= \quad C_1^1 + C_2^1 + C_3^1 + \cdots + C_n^1$$
$$= \quad C_{n+1}^2$$
$$= \quad \frac{n(n+1)}{2}$$

ⓘ *Tip: More explanation will be given in the next lecture.*

---

4) Evaluate $1 \times 2 + 2 \times 3 + 3 \times 4 + \cdots + n(n+1)$.

$$1 \times 2 + 2 \times 3 + 3 \times 4 + \cdots + n(n+1)$$
$$-2! \times \left(\frac{1 \times 2}{2!} + \frac{2 \times 3}{2!} + \cdots + \frac{n(n+1)}{2!} \right)$$
$$= 2! \times \left(C_2^2 + C_3^2 + \cdots + C_{n+1}^2 \right)$$

$$=2! \times C_{n+2}^3$$

$$=\frac{(n+2)(n+1)n}{3}$$

ⓘ   *Tip: More explanation will be given in the next lecture.*

---

5) Compute $1 \times 2 \times 3 + 2 \times 3 \times 4 + \cdots + 2016 \times 2017 \times 2018$.

$$1 \times 2 \times 3 + 2 \times 3 \times 4 + \cdots + 2016 \times 2017 \times 2018$$

$$=3! \times \left(\frac{1 \times 2 \times 3}{3!} + \frac{2 \times 3 \times 4}{3!} + \cdots + \frac{2016 \times 2017 \times 2018}{3!} \right)$$

$$=3! \times \left(C_3^3 + C_4^3 + \cdots + C_{2018}^3 \right)$$

$$=6 \times C_{2019}^4$$

Lecture 12

1) Simplify $1 + 3 + 6 + \cdots + \frac{n(n+1)}{2}$.

Solution 1

$$1 + 3 + 6 + \cdots + \frac{n(n+1)}{2}$$

$$= \sum_{k=1}^{n} \left(\frac{k(k+1)}{2} \right)$$

$$= \frac{1}{2} \sum_{k=1}^{n} (k^2 + k)$$

$$= \frac{1}{2} \left(\sum_{k=1}^{n} k^2 + \sum_{k=1}^{n} k \right)$$

$$= \frac{1}{2} \left(\frac{n(n+1)(2n+1)}{6} + \frac{n(n+1)}{2} \right)$$

$$= \frac{1}{2} \times \frac{n(n+1)(2n+1+3)}{6}$$

$$= \frac{n(n+1)(n+2)}{6}$$

Solution 2

$$1 + 3 + 6 + \cdots + \frac{n(n+1)}{2}$$

$$= C_2^2 + C_3^2 + C_4^2 + \cdots + C_{n+1}^2$$

$$= C_{n+2}^3$$

$$= \frac{n(n+1)(n+2)}{6}$$

2) Explain the principle of mathematical induction.

$$\boxed{n=1} \longrightarrow \boxed{n=2} \rightarrow \cdots \rightarrow \boxed{n=k} \longrightarrow \boxed{n=k+1} \cdots$$

In order to prove a claim holds for all positive integers, it is sufficient to show that

(a) the claim holds when $n = 1$, i.e. the initial condition

(b) if the claim holds when $n = k$, then it must hold when $n = k+1$, i.e. the recursion

This is because, from (a) above (the claim holds when $n = 1$), we can conclude the claim will hold for $n = 2$ by using (b) above. Then, by the same reasoning, we can conclude that the claim holds when $n = 3, 4, \cdots$. This derivation is repeatable and can last forever. Therefore the claim must hold for all positive integers.

> There are some variations of the mathematical induction. But the core principle is similar, if not exactly the same.

3) Using the principle of mathematical induction to prove

$$1^3 + 2^3 + \cdots + n^3 = \left[\frac{n(n+1)}{2}\right]^2 = (1 + 2 + \cdots + n)^2$$

When $n = 1$, the left side equals 1 and the right side is also 1. Therefore the claim holds.

Suppose the claim holds when $n = k$, or

$$1^3 + 2^3 + \cdots + k^3 = \left[\frac{k(k+1)}{2}\right]^2$$

Then when $n = k + 1$:

$$1^3 + 2^3 + \cdots + k^3 + (k+1)^3$$

$$=\left[\frac{k(k+1)}{2}\right]^2 + (k+1)^3$$

$$=\frac{1}{4} \times \left[k^2(k+1)^2 + 4(k+1)^3\right]$$

$$=\frac{1}{4} \times (k+1)^2 \times (k^2 + 4(k+1))$$

$$=\frac{1}{4} \times (k+1)^2 \times (k+2)^2$$

$$=\left[\frac{(k+1)(k+2)}{2}\right]^2$$

Therefore by the principle of mathematical induction, the claim holds for all positive integer n.

4) Show that $1(1!) + 2(2!) + 3(3!) + \cdots + n(n!) = (n+1)! - 1$

This claim can be proved using mathematical induction.

When $n = 1$, the left side equals $1(1!) = 1$ and the right side equals $(1+1)! - 1 = 1$. Apparently, the claim holds.

Suppose the claim holds when $n = k$, i.e.

$$1(1!) + 2(2!) + \cdots + k(k!) = (k+1)! - 1$$

Then when $n = k+1$,

$$1(1!) + 2(2!) + \cdots + k(k!) + (k+1)((k+1)!)$$
$$=(k+1)! - 1 + (k+1)((k+1)!)$$
$$=(k+1)!(1 + (k+1)) - 1$$
$$=(k+2)! - 1$$

The claim still holds when $n = k+1$.

Hence by the principle of mathematical induction, this claim holds for all positive integers.

5) Simplify $3^3 + 6^3 + 9^3 + \cdots + (3n)^3$.

$$3^3 + 6^3 + 9^3 + \cdots + (3n)^3$$
$$=27 \times (1^3 + 2^3 + \cdots + n^3)$$
$$=27 \times \left[\frac{n(n+1)}{2}\right]^2$$
$$=\frac{27}{4} \times \left[n(n+1)\right]^2$$

6) Find the sum formula $1^3 + 2^3 + \cdots + n^3$ directly.

Because $k^3 = k(k+1)(k+2) - 3k^2 - 2k$, we find

$$1^3 + 2^3 + \cdots + n^3$$
$$=(1 \times 2 \times 3 - 3 \times 1^2 - 2 \times 1)+$$
$$(2 \times 3 \times 4 - 3 \times 2^2 - 2 \times 2)+$$
$$\cdots +$$
$$(n \times (n+1) \times (n+2) - 3 \times n^2 - 2 \times n)$$
$$=(1 \times 2 \times 3 + 2 \times 3 \times 4 + \cdots + n(n+1)(n+2))-$$
$$3 \times (1^2 + 2^2 + \cdots + n^2)-$$
$$2 \times (1 + 2 + \cdots + n)$$
$$=\frac{1}{4}n(n+1)(n+2)(n+3) - 3 \times \frac{1}{6}n(n+1)(2n+1) - 2 \times \frac{1}{2}n(n+1)$$
$$=\frac{1}{4}n(n+1)\left((n+2)(n+3) - 2(2n+1) - 4\right)$$
$$=\frac{1}{4}n^2(n+1)^2$$

7) Simplify $1^2 + 3^2 + \cdots + (2n-1)^2$.

Solution 1

$$\sum_{k=1}^{n} (2k-1)^2$$

$$= \sum_{k=1}^{n} (4k^2 - 4k + 1)$$

$$= 4\sum_{k=1}^{n} k^2 - 4\sum_{k=1}^{n} k + n$$

$$= 4 \times \frac{n(n+1)(2n+1)}{6} - 4 \times \frac{n(n+1)}{2} + n$$

$$= \frac{n(2n-1)(2n+1)}{3}$$

Solution 2

$$1^2 + 3^2 + \cdots + (2n-1)^2$$

$$= (1^2 + 2^2 + \cdots + (2n-1)^2) - (2^2 + 4^2 + \cdots + (2n-2)^2)$$

$$= \frac{1}{6}(2n-1)((2n-1)+1)(2(2n-1)+1)$$

$$\quad - 4 \times (1^2 + 2^2 + \cdots + (n-1)^2)$$

$$= \frac{1}{6}(2n-1)(2n)(4n-1) - 4 \times \frac{1}{6}(n-1)(n)(2n-1)$$

$$= \frac{n(2n-1)(2n+1)}{3}$$

8) Let n be a positive integer. Show that

$$\left(1 + \frac{1}{3}\right)\left(1 + \frac{1}{3^2}\right) \cdots \left(1 + \frac{1}{3^n}\right) < 2$$

(Ref 2012 China)

Let $f(n) = \left(1 + \frac{1}{3}\right)\left(1 + \frac{1}{3^2}\right) \cdots \left(1 + \frac{1}{3^n}\right)$.

Instead showing $f(n) < 2$, let's prove $f(n) < 2 - \frac{1}{3^n}$ using mathematical induction.

When $n = 1$, $f(1) = 1 + \frac{1}{3} < 2 - \frac{1}{3}$. The claim holds.

Suppose the claim holds when $n = k$, i.e. $f(k) < 2 - \frac{1}{3^k}$. Then when $n = k + 1$,

$$f(k+1) = f(k)\left(1 + \frac{1}{3^{k+1}}\right)$$
$$< \left(2 - \frac{1}{3^k}\right)\left(1 + \frac{1}{3^{k+1}}\right)$$
$$= 2 - \frac{1}{3^k} + \frac{2}{3^{k+1}} - \frac{1}{3^{2k+1}}$$
$$< 2 - \frac{1}{3^k} + \frac{2}{3^{k+1}}$$
$$= 2 - \frac{1}{3^{k+1}}$$

Therefore, by the principle of mathematical principle, it always hold that
$$f(n) < 2 - \frac{1}{3^n} < 2$$

> Tip: Mathematical induction is always a good try to solve claims related to positive integers.

Lecture 13

1. Let m, k, and n be 3 positive integers satisfying $m \leq k \leq n$. Show that

$$C_n^k C_k^m = C_n^m C_{n-m}^{k-m} = C_n^{k-m} C_{n-k+m}^k \qquad (15.7)$$

Let's prove this claim using basic definitions.

$$
\begin{aligned}
C_n^k C_k^m &= \frac{n!}{k!(n-k)!} \cdot \frac{k!}{m!(k-m)!} &= \frac{n!}{m!(n-k)!(k-m)!} \\
C_n^m C_{n-m}^{k-m} &= \frac{n!}{m!(n-m)!} \cdot \frac{(n-m)!}{(k-m)!(n-k)!} &= \frac{n!}{m!(n-k)!(k-m)!} \\
C_n^{k-m} C_{n-k+m}^m &= \frac{n!}{(k-m)!(n-k+m)!} \cdot \frac{(n-k+m)!}{m!(n-k)!} &= \frac{n!}{m!(n-k)!(k-m)!}
\end{aligned}
$$

Therefore these three terms are equal.

2. Show that

$$\frac{1}{C_n^k} = \frac{k}{k-1}\left(\frac{1}{C_{n-1}^{k-1}} - \frac{1}{C_n^{k-1}}\right) \qquad (15.8)$$

This identity can be proved using basic definitions. From the right side

$$
\begin{aligned}
&\frac{k}{k-1}\left(\frac{1}{C_{n-1}^{k-1}} - \frac{1}{C_n^{k-1}}\right) \\
&= \frac{k}{k-1}\left(\frac{(k-1)!(n-k)!}{(n-1)!} - \frac{(k-1)!(n-k+1)!}{n!}\right) \\
&= \frac{k}{k-1} \cdot \frac{n(k-1)!(n-k)! - (k-1)!(n-k+1)!}{n!} \\
&= \frac{k!(n(n-k)! - (n-k+1)!)}{(k-1)n!} \\
&= \frac{k!(n-k)!(n - (n-k+1))}{(k-1)n!}
\end{aligned}
$$

$$= \frac{k!(n-k)!}{n!}$$

$$= \frac{1}{C_n^k}$$

3. Simplify

$$\sum_{k=1}^{n} k^2 C_n^k$$

$$\sum_{k=1}^{n} k^2 C_n^k$$

$$= \sum_{k=1}^{n} kn C_{n-1}^{k-1} \qquad \qquad \because \text{ by } \textit{Formula 13.2}$$

$$= n \sum_{k=1}^{n} (k-1+1) C_{n-1}^{k-1}$$

$$= n \left(\sum_{k=1}^{n} (k-1) C_{n-1}^{k-1} + \sum_{k=1}^{n} C_{n-1}^{k-1} \right)$$

$$= n \sum_{k=0}^{n-1} k C_{n-1}^{k} + n \sum_{k=0}^{n-1} C_{n-1}^{k} \qquad \qquad \text{replace subscripts}$$

$$= n \sum_{k=1}^{n-1} (n-1) C_{n-2}^{k-1} + n \times 2^{n-1} \qquad \qquad \because \text{ by } \textit{Formula 13.2}$$

$$= n(n-1) \sum_{k=1}^{n-1} C_{n-2}^{k-1} + n \times 2^{n-1}$$

$$= n(n-1) \times 2^{n-2} + n \times 2^{n-1}$$

$$= n(n+1) \times 2^{n-2}$$

4. Let positive integers m and n satisfy $m \leq n$. Prove

$$\sum_{k=m}^{n} C_n^k C_k^m = 2^{n-m} C_n^m$$

By *Equation 15.7* on *page 123*

$$\sum_{k=m}^{n} C_n^k C_k^m = C_n^m \sum_{k=m}^{n} C_{n-m}^{k-m} = C_n^m \sum_{l=0}^{n-m} C_{n-m}^{l} = 2^{n-m} C_n^m$$

5. Show that

$$C_n^1 - \frac{1}{2}C_n^2 + \frac{1}{3}C_n^3 - \cdots + (-1)^{n+1}C_n^n = 1 + \frac{1}{2} + \frac{1}{3} + \cdots + \frac{1}{n}$$

First, by *Equation (13.1)* and *(13.2)*, we have

$$\frac{1}{k}C_n^k = \frac{1}{k}\left(C_{n-1}^k + C_{n-1}^{k-1}\right) = \frac{1}{k}C_{n-1}^k + \frac{1}{n}C_n^k$$

Let $f(n)$ be the left side of the to-be-proved identity, we are going to show the following recursion holds

$$f(n) = f(n-1) + \frac{1}{n}$$

This is because

$$
\begin{aligned}
f(n) =& C_n^1 - \frac{1}{2}C_n^2 + \cdots + (-1)^n \frac{1}{n-1}C_n^{n-1} + (-1)^{n+1}\frac{1}{n}C_n^n \\
=& \left(C_{n-1}^1 + \frac{1}{n}C_n^1\right) - \left(\frac{1}{2}C_{n-1}^2 + \frac{1}{n}C_n^2\right) + \cdots \\
&+ (-1)^n\left(\frac{1}{n-1}C_{n-1}^{n-1} + \frac{1}{n}C_n^{n-1}\right) + (-1)^{n+1}\left(\frac{1}{n}\right) \\
=& \left(C_{n-1}^1 - \frac{1}{2}C_{n-1}^2 + \cdots + (-1)^n\frac{1}{n-1}C_{n-1}^{n-1}\right) \\
&+ \frac{1}{n}(C_n^1 - C_n^2 + \cdots + (-1)^{n+1}C_n^n) \\
=& f(n-1) + \frac{1}{n} \qquad \text{by *practice 15.6* on *page 110*}
\end{aligned}
$$

Therefore we have

$$f(n) = f(n-1) + \frac{1}{n}$$

$$= f(n-2) + \frac{1}{n-1} + \frac{1}{n}$$

$$\cdots$$

$$= f(1) + \frac{1}{2} + \cdots + \frac{1}{n-1} + \frac{1}{n}$$

$$= 1 + \frac{1}{2} + \cdots + \frac{1}{n-1} + \frac{1}{n}$$

6. Show that

$$\frac{1}{C_{2n}^0} - \frac{2}{C_{2n}^1} + \cdots - \frac{2n}{C_{2n}^{2n-1}} = 0$$

By *Equation 13.2* on *page 55*:

$$\frac{k+1}{C_{2n}^k} = \frac{2n+1}{C_{2n+1}^{k+1}} = \frac{2n+1}{C_{2n+1}^{2n-k}} = \frac{2n-k}{C_{2n}^{2n-k-1}}$$

Therefore, we have

$$k = 0 \implies \frac{1}{C_{2n}^0} = \frac{2n}{C_{2n}^{2n-1}}$$

$$k = 1 \implies \frac{2}{C_{2n}^1} = \frac{2n-1}{C_{2n}^{2n-2}}$$

$$\cdots$$

$$k = n-1 \implies \frac{n}{C_{2n}^{n-1}} = \frac{n+1}{C_{2n}^n}$$

Now it is clearly that all the terms in the left side of the given expression will be canceled in pair. Therefore their sum must equal 0.

7. Evaluate

$$\sum_{n=2016}^{\infty} \frac{1}{C_n^{2016}}$$

By *Equation 15.8* on *page 123*, we have

$$\frac{1}{C_n^{2016}} = \frac{2016}{2015}\left(\frac{1}{C_{n-1}^{2015}} - \frac{1}{C_n^{2015}}\right)$$

$$\sum_{n=2016}^{\infty} \frac{1}{C_n^{2016}} = \frac{2016}{2015}\left[\left(\frac{1}{C_{2015}^{2015}} - \frac{1}{C_{2016}^{2015}}\right) + \left(\frac{1}{C_{2016}^{2015}} - \frac{1}{C_{2017}^{2015}}\right) + \cdots\right]$$

$$= \frac{2016}{2015}\left(\frac{1}{C_{2015}^{2015}} - \frac{1}{C_{\infty}^{2015}}\right)$$

$$= \boxed{\frac{2016}{2015}}$$

Lecture 14

1) Find the value of $\cos 20° \cos 40° \cos 80°$.

Let the desired result be S and multiply it with $\sin 20°$:

$$\sin 20° \cdot S = \sin 20° \cos 20° \cos 40° \cos 80°$$

$$= \frac{1}{2} \cdot \sin 40° \cos 40° \cos 80°$$

$$= \frac{1}{4} \cdot \sin 80° \cos 80°$$

$$= \frac{1}{8} \cdot \sin 160°$$

$$= \frac{1}{8} \cdot \sin 20°$$

This means

$$\sin 20° \cdot S = \frac{1}{8} \cdot \sin 20° \implies S = \boxed{\frac{1}{8}}$$

---

2) Let $\alpha \in \left(\frac{3\pi}{2}, 2\pi \right)$. Simplify

$$\sqrt{ \frac{1}{2} - \frac{1}{2} \sqrt{ \frac{1}{2} + \frac{1}{2} \cdot \cos 2\alpha } }$$

$$\alpha \in \left(\frac{3\pi}{2}, 2\pi \right) \implies \sqrt{ \frac{1}{2} + \frac{1}{2} \cdot \cos 2\alpha } = |\cos \alpha| = \cos \alpha$$

also

$$\sqrt{ \frac{1}{2} - \frac{1}{2} \cdot \cos \alpha } = \left| \sin \frac{\alpha}{2} \right| = \sin \frac{\alpha}{2}$$

Hence the final result is $\boxed{\sin \frac{\alpha}{2}}$.

3) Simplify $\cos x \cos 2x \cdots \cos 2^{n-1}x$.

$$\cos x \cos 2x \cdots \cos 2^{n-1}x$$
$$= \frac{\sin 2x}{2 \sin x} \cdot \frac{\sin 4x}{2 \sin 2x} \cdots \frac{\sin 2^n x}{2 \sin 2^{n-1}x}$$
$$= \boxed{\frac{\sin 2^n x}{2^n \sin x}}$$

4) Show that

$$\sin x + 2 \sin 2x + \cdots + n \sin nx = \frac{(n+1) \sin nx - n \sin (n+1)x}{2(1 - \cos x)}$$

Let $S = \sin x + 2 \sin 2x + \cdots + n \sin nx$. Then

$2 \cos x \cdot S$
$= \sin 2x + 2(\sin 3x + \sin x) + \cdots + n(\sin (n+1)x + \sin (n-1)x)$
$= 2 \cdot S + n \sin (n+1)x - (n+1) \sin nx$

$$\therefore \quad S = \frac{(n+1) \sin nx - n \sin (n+1)x}{2(1 - \cos x)}$$

5) Compute the value of

$$\cos \frac{\pi}{2n+1} \cdot \cos \frac{2\pi}{2n+1} \cos \frac{3\pi}{2n+1} \cdots \cos \frac{n\pi}{2n+1}$$

Let

$$A = \cos\frac{\pi}{2n+1} \cdot \cos\frac{2\pi}{2n+1} \cdots \cos\frac{n\pi}{2n+1}$$

$$B = \sin\frac{\pi}{2n+1} \cdot \sin\frac{2\pi}{2n+1} \cdots \sin\frac{n\pi}{2n+1}$$

Then

$$A \cdot B$$

$$= \left(\frac{1}{2}\sin\frac{2\pi}{2n+1}\right)\left(\frac{1}{2}\sin\frac{4\pi}{2n+1}\right)\left(\frac{1}{2}\sin\frac{6\pi}{2n+1}\right)\cdots\left(\frac{1}{2}\sin\frac{2n\pi}{2n+1}\right)$$

$$= \frac{1}{2^n}\sin\frac{2\pi}{2n+1}\sin\frac{4\pi}{2n+1}\sin\frac{6\pi}{2n+1}\cdots\sin\frac{3\pi}{2n+1}\sin\frac{\pi}{2n+1}$$

$$= \frac{1}{2^n}B$$

$$\therefore \quad AB = \frac{1}{2^n}B \implies A = \boxed{\frac{1}{2^n}}$$

6) Evaluate

$$\cos\frac{2\pi}{2n+1} + \cos\frac{4\pi}{2n+1} + \cdots + \cos\frac{2n\pi}{2n+1}$$

Let $\theta = \frac{\pi}{2n+1} \implies \sin(2n+1)\theta = 0$. Then the desired sum equals

$$S = \cos 2\theta + \cos 4\theta + \cdots + \cos 2n\theta$$

Multiplying S by $2\sin\theta$:

$$2\sin\theta\cos 2\theta = \sin 3\theta - \sin\theta$$
$$2\sin\theta\cos 4\theta = \sin 5\theta - \sin 3\theta$$
$$\cdots$$
$$2\sin\theta\cos 2n\theta = \sin(2n+1)\theta - \sin(2n-1)\theta$$

Adding these equations yields

$$2\sin\theta \cdot S = \sin(2n+1)\theta - \sin\theta = -\sin\theta \implies S = \boxed{-\frac{1}{2}}$$

7) Evaluate

$$\cos \frac{\pi}{2n+1} + \cos \frac{3\pi}{2n+1} + \cdots + \cos \frac{(2n-1)\pi}{2n+1}$$

$$\cos \frac{\pi}{2n+1} + \cos \frac{3\pi}{2n+1} + \cdots + \cos \frac{(2n-1)\pi}{2n+1}$$

$$= -\left(\cos \frac{2n\pi}{2n+1} + \cos \frac{(2n-2)\pi}{2n+1} + \cdots + \cos \frac{2\pi}{2n+1} \right)$$

$$= \boxed{\frac{1}{2}} \qquad \text{(by the previous practice)}$$

8) Without using a calculator, find the value of

$$\cos \frac{\pi}{13} + \cos \frac{3\pi}{13} + \cos \frac{9\pi}{13}$$

Let

$$x = \cos \frac{\pi}{13} + \cos \frac{3\pi}{13} + \frac{9\pi}{13}$$

and

$$y = \cos \frac{5\pi}{13} + \cos \frac{7\pi}{13} + \frac{11\pi}{13}$$

Then by the previous practice, we find

$$x + y = \frac{\pi}{13} + \frac{3\pi}{13} + \frac{5\pi}{13} + \frac{7\pi}{13} + \frac{9\pi}{13} + \frac{11\pi}{13} = \frac{1}{2}$$

and

$$xy = \left(\cos \frac{\pi}{13} + \cos \frac{3\pi}{13} + \frac{9\pi}{13} \right)\left(\cos \frac{5\pi}{13} + \cos \frac{7\pi}{13} + \frac{11\pi}{13} \right)$$

$$= -\frac{3}{2} \cdot \left(\cos \frac{\pi}{13} - \cos \frac{2\pi}{13} + \cos \frac{3\pi}{13} - \cos \frac{4\pi}{13} + \cos \frac{5\pi}{13} - \cos \frac{6\pi}{13} \right)$$

$$= -\frac{3}{2} \cdot \left(\cos \frac{\pi}{13} + \cos \frac{3\pi}{13} + \cos \frac{5\pi}{13} + \cos \frac{7\pi}{13} + \cos \frac{9\pi}{13} + \cos \frac{11\pi}{13} \right)$$

$$= -\frac{3}{4}$$

Therefore by Vieta's theorem, x and y are the two roots of the following equation

$$u^2 - \frac{1}{2} \cdot u - \frac{3}{4} = 0$$

Because apparently, $x > 0$, we find $x = \boxed{\dfrac{1 + \sqrt{13}}{4}}$.

---

9) Evaluate

$$(1 + \tan 1°)(1 + \tan 2°) \cdots (1 + \tan 44°)(1 + \tan 45°)$$

Note that

$$(1 + \tan \alpha)(1 + \tan(45° - \alpha)) = (1 + \tan \alpha)\left(1 + \frac{\tan 45° - \tan \alpha}{1 + \tan 45° \tan \alpha}\right) = 2$$

We find the original expression equals $\boxed{2^{23}}$.

---

10) Evaluate

$$\cos \frac{\pi}{2n + 1} \cdot \cos \frac{2\pi}{2n + 1} \cdot \cos \frac{3\pi}{2n + 1} \cdots \cos \frac{n\pi}{2n + 1}$$

Let

$$C = \cos \frac{\pi}{2n+1} \cdot \cos \frac{2\pi}{2n+1} \cdot \cos \frac{3\pi}{2n+1} \cdots \cos \frac{n\pi}{2n+1}$$

and

$$S = \sin \frac{\pi}{2n+1} \cdot \sin \frac{2\pi}{2n+1} \cdot \sin \frac{3\pi}{2n+1} \cdots \sin \frac{n\pi}{2n+1}$$

Then

$$C \cdot S = \frac{1}{2^n} \cdot \sin \frac{2\pi}{2n+1} \cdot \sin \frac{4\pi}{2n+1} \cdots \sin \frac{2(n-1)\pi}{2n+1} \sin \frac{2n\pi}{2n+1}$$
$$= \frac{1}{2^n} \cdot \sin \frac{2\pi}{2n+1} \cdot \sin \frac{4\pi}{2n+1} \cdots \sin \frac{3\pi}{2n+1} \cdot \sin \frac{\pi}{2n+1}$$
$$= \frac{1}{2^n} \cdot S$$

Hence, we conclude $\boxed{C = \dfrac{1}{2^n}}$.

---

11) Let sequence $\{a_n\}$ satisfy the condition: $a_1 = \frac{\pi}{6}$ and $a_{n+1} = \arctan(\sec a_n)$, where $n \in Z^+$. There exists a positive integer m such that $\sin a_1 \cdot \sin a_2 \cdots \sin a_m = \frac{1}{100}$. Find m.
(Ref 2014 China)

From the given information, it is apparent that $a_{n+1} \in (-\frac{\pi}{2}, \frac{\pi}{2})$ holds for every $n \in Z^+$, and

$$\tan a_{n+1} = \sec a_n \tag{15.9}$$

Because $\sec a_1 > 0$, we can conclude $a_{n+1} \in (0, \frac{\pi}{2})$ for every n.
By (15.9), we have $\tan^2 a_{n+1} = \sec^2 a_n = 1 + \tan^2 a_n$. Or

$$\tan^2 a_n = (n-1) + \tan^2 a_1 = (n-1) + \frac{1}{3} \implies \tan a_n = \frac{3n-2}{3}$$

Solution

Therefore:

$$
\begin{aligned}
&\sin a_1 \cdot \sin a_2 \cdots \sin a_m \\
&= \frac{\tan a_1}{\sec a_1} \cdot \frac{\tan a_2}{\sec a_2} \cdots \frac{\tan a_m}{\sec a_m} \\
&= \frac{\tan a_1}{\tan a_2} \cdot \frac{\tan a_2}{\tan a_3} \cdots \frac{\tan a_m}{\tan a_{m+1}} \\
&= \frac{\tan a_1}{\tan a_{m+1}} \\
&= \sqrt{\frac{1}{3m+1}}
\end{aligned}
$$

Setting

$$
\sqrt{\frac{1}{3m+1}} = \frac{1}{100}
$$

yields $m = 3333$.

Lecture 15

1) Use *Equation 15.2* on *page 65* to prove $\sin^2\theta + \cos^2\theta = 1$.

Let $z = \cos\theta + i\sin\theta$, then

$$\sin^2\theta + \cos^2\theta$$
$$= \left[\frac{1}{2} \times \left(z - \frac{1}{z}\right)i\right]^2 + \left[\frac{1}{2} \times \left(z + \frac{1}{z}\right)\right]^2$$
$$= \frac{1}{4} \times \left[-\left(z^2 - 2 + \frac{1}{z^2}\right) + \left(z^2 + 2 + \frac{1}{z^2}\right)\right]$$
$$= 1$$

2) Let sequences a_n and b_n satisfy $a_n = a_{n-1}\cos\theta - b_{n-1}\sin\theta$ and $b_n = a_{n-1}\sin\theta + b_{n-1}\cos\theta$. If $a_1 = 1$ and $b_1 = \tan\theta$, where θ is a known real number, find the general formulas for a_n and b_n

Let $z_n = a_n + i \cdot b_n$, then

$$\frac{z_n}{z_{n-1}} = \frac{(a_{n-1}\cos\theta - b_{n-1}\sin\theta) + i \cdot (a_{n-1}\sin\theta + b_{n-1}\cos\theta)}{a_{n-1} + i \cdot b_{n-1}}$$
$$= \cos\theta + i \cdot \sin\theta$$

Hence, z_n is a geometric sequence.

$$z_n = (1 + i \cdot \tan\theta)(\cos\theta + i \cdot \sin\theta)^{n-1}$$
$$= \sec\theta(\cos\theta + i \cdot \sin\theta)^n$$
$$= \sec\theta\cos n\theta + i \cdot \sin n\theta$$

This implies

$$a_n = \sec\theta\cos n\theta$$
$$b_n = \sec\theta\sin n\theta$$

3) Simplify

$$\sin x + \sin 2x + \cdots + \sin nx$$

and

$$\cos x + \cos 2x + \cdots + \cos nx$$

Let $z = \cos x + i \sin x$, then

$$\cos kx = \frac{1}{2} \times \left(z^k + \frac{1}{z^k} \right) \quad \text{and} \quad \sin kx = -\frac{1}{2} \times \left(z^k - \frac{1}{z^k} \right) i$$

It follows that

$$\cos x + \cos 2x + \cdots + \cos nx$$

$$= \frac{1}{2} \times \left(z + \frac{1}{z} \right) + \frac{1}{2} \times \left(z^2 + \frac{1}{z^2} \right) + \cdots + \frac{1}{2} \times \left(z^n + \frac{1}{z^n} \right)$$

$$= \frac{1}{2} \left(z + z^2 + \cdots + z^n \right) + \frac{1}{2} \left(\frac{1}{z} + \frac{1}{z^2} + \cdots + \frac{1}{z^n} \right)$$

$$= \frac{1}{2} \times z \times \frac{1 - z^n}{1 - z} + \frac{1}{2} \times \frac{1}{z} \times \frac{1 - \frac{1}{z^n}}{1 - \frac{1}{z}}$$

$$= \frac{1}{2} \times \frac{(z^n - 1)(z^{n+1} + 1)}{z^n(z - 1)}$$

In order to convert it back to trigonometric expression, we need to transform every term to the form of $\left(x^k \pm \frac{1}{x^k} \right)$.

Dividing both the denominator and numerator part by $z^{n+\frac{1}{2}}$ yields

$$\frac{1}{2} \times \frac{\left(z^{\frac{n}{2}} - \frac{1}{z^{\frac{n}{2}}} \right)\left(z^{\frac{n+1}{2}} + \frac{1}{z^{\frac{n+1}{2}}} \right)}{z^{\frac{1}{2}} - \frac{1}{z^{\frac{1}{2}}}} = \boxed{\frac{\sin\left(\frac{nx}{2}\right) \cos\left(\frac{(n+1)x}{2}\right)}{\sin\left(\frac{x}{2}\right)}}$$

Using a similar method, we have

$$\sin x + \sin 2x + \cdots + \sin nx$$

$$= -\frac{i}{2}\left[\left(z - \frac{1}{z}\right) + \left(z^2 - \frac{1}{z^2}\right) + \cdots + \left(z^n - \frac{1}{z^n}\right)\right]$$

$$= -\frac{i}{2}\left[\left(z + z^2 + \cdots + z^n\right) - \left(\frac{1}{z} + \frac{1}{z^2} + \cdots + \frac{1}{z^n}\right)\right]$$

$$= -\frac{i}{2} \times \left[z \times \frac{1 - z^n}{1 - z} - \frac{1}{z} \times \frac{1 - \frac{1}{z^n}}{1 - \frac{1}{z}}\right]$$

$$= -\frac{i}{2} \times \frac{(z^n - 1)(z^{n+1} - 1)}{z^n(z - 1)}$$

$$= -\frac{i}{2} \times \frac{(z^{\frac{n}{2}} - \frac{1}{z^{\frac{n}{2}}})(z^{\frac{n+1}{2}} - \frac{1}{z^{\frac{n+1}{2}}})}{z^{\frac{1}{2}} - \frac{1}{z^{\frac{1}{2}}}}$$

$$= \boxed{\frac{\sin(\frac{nx}{2})\sin(\frac{(n+1)x}{2})}{\sin(\frac{x}{2})}}$$

ⓘ *Tip: The result suggests that this problem can also be solved by multiplying these two expressions by $\sin(\frac{x}{2})$ first and then applying the product to sum formula.*

4) Solve the equation $\cos\theta + \cos 2\theta + \cos 3\theta = \sin\theta + \sin 2\theta + \sin 3\theta$.

Let $z = \cos\theta + i\sin\theta$, then the give equation is equivalent to

$$\frac{1}{2} \times \left((z + \frac{1}{z}) + (z^2 + \frac{1}{z^2}) + (z^3 + \frac{1}{z^3})\right)$$

$$= -\frac{i}{2} \times \left((z - \frac{1}{z}) + (z^2 - \frac{1}{z^2}) + (z^3 - \frac{1}{z^3})\right)$$

or

$$(z^6 + z^5 + z^4)(1 + i) + (z^2 + z + 1)(1 - i) = 0$$
$$(z^2 + z + 1)(z^4(1 + i) + (1 - i)) - 0$$
$$(z^2 + z + 1)(z^4 - i) = 0$$

Therefore

$$z^2 + z + 1 = 0 \implies \theta = \boxed{\left((2k+1) \pm \frac{1}{3}\right)\pi}$$

$$z^4 - i = 0 \implies \theta = \boxed{\left(\frac{k}{2} + \frac{1}{8}\right)\pi}$$

where k is an integer.

5) Solve the equation $\cos^2 x + \cos^2 2x + \cos^2 3x = 1$ in $(0, 2\pi)$.

(Ref 1962 IMO)

Let $z = \cos x + i \sin x$. Then

$$z^2 + \frac{1}{z^2} + z^4 + \frac{1}{z^4} + z^6 + \frac{1}{z^6} = -2$$

Substituting $w = z^2 + \frac{1}{z^2}$ and simplifying yields:

$$w^3 + w^2 - 2w = 0 \implies w(w+2)(w-1) = 0 \implies w = 0, 1, -2$$

When $w = 0 \implies z^2 + \frac{1}{z^2} = 0 \implies z^4 = 1 \implies x = 45°, 135°, 225°, 315°$.

When $w = 1 \implies z^2 + \frac{1}{z^2} = 1 \implies z^4 - z^2 + 1 = 0 \implies z^2 = \cos 60° + i \sin 60°$, or $\cos 300° + i \sin 300° \implies x = 30°, 150°, 210°, 330°$.

When $w = -2 \implies z^2 + \frac{1}{z^2} = -2 \implies (z^2 - 1)^2 = 0 \implies z = \pm 1 \implies x = 180°$.

Hence we conclude

$$x = \boxed{30°, 45°, 135°, 150°, 180°, 210°, 225°, 315°, 330°}$$

6) Let A, B, and C be a triangle's three inner angles. Let $a = \cos A + i \sin A$, $b = \cos B + i \sin B$, and $c = \cos C + i \sin C$. Show that

$$abc = -1 \qquad (15.10)$$

$$
\begin{aligned}
abc &= (\cos A + i \sin A)(\cos B + i \sin B)(\cos C + i \sin C) \\
&= \cos(A + B + C) + i \sin(A + B + C) \\
&= \cos \pi + i \sin \pi \\
&= -1
\end{aligned}
$$

7) Let A, B, and C be angles of a triangle. If $\cos 3A + \cos 3B + \cos 3C = 1$, determine the largest angle of the triangle.

Let $a = \cos A + i \sin A$, $b = \cos B + i \sin B$, and $c = \cos C + i \sin C$. Then

$$
\begin{aligned}
&\cos 3A + \cos 3B + \cos 3C = 1 \\
\Leftrightarrow\ & a^3 + \frac{1}{a^3} + b^3 + \frac{1}{b^3} + c^3 + \frac{1}{c^3} = 2 \\
\Leftrightarrow\ & a^3 - b^3 c^3 + b^3 - c^3 a^3 + c^3 - a^3 b^3 = 2 \qquad \because (15.10) \\
\Leftrightarrow\ & a^3 + b^3 + c^3 - a^3 b^3 - b^3 c^3 - c^3 a^3 = 1 - a^3 b^3 c^3 \\
\Leftrightarrow\ & (a^3 - 1)(b^3 - 1)(c^3 - 1) = 0
\end{aligned}
$$

This implies one of a, b, and c must be $\cos \frac{2\pi}{3} + i \sin \frac{2\pi}{3}$. Therefore we conclude that the largest angle must be $\boxed{\dfrac{2\pi}{3}}$ or $\boxed{120°}$.

8) Evaluate $\sin \theta + \frac{1}{2} \cdot \sin 2\theta + \frac{1}{4} \cdot \sin 3\theta + \cdots$.

Solution

Let $z = \cos\theta + i\sin\theta$. Then

$$\left(\cos\theta + \frac{1}{2}\cdot\cos 2\theta + \frac{1}{4}\cdot\cos 3\theta\right) + i\left(\sin\theta + \frac{1}{2}\cdot\theta + \frac{1}{4}\cdot 3\theta + \cdots\right)$$

$$= z + \frac{1}{2}\cdot z^2 + \frac{1}{4}\cdot z^3 + \cdots$$

$$= \frac{z}{1 - \frac{1}{2}\cdot z}$$

$$= \frac{2z}{2 - z}$$

$$= \frac{2(\cos\theta + i\sin\theta)}{(2 - \cos\theta) - i\sin\theta}$$

$$= \frac{2\left(\cos\theta + i\sin\theta\right)\left((2 - \cos\theta) + i\sin\theta\right)}{\left((2 - \cos\theta) - i\sin\theta\right)\left((2 - \cos\theta) + i\sin\theta\right)}$$

$$= \frac{(\ldots) + i(4\sin\theta)}{5 - 4\cos\theta}$$

Where $(\cdots)$ is an expression of θ which does not involve i.
Therefore we conclude

$$\sin\theta + \frac{1}{2}\cdot\sin 2\theta + \frac{1}{4}\cdot\sin 3\theta + \cdots = \boxed{\frac{4\sin\theta}{5 - 4\cos\theta}}$$

9) Prove for every positive integer n and real number $x \neq \frac{k\pi}{2^t}$ where $t = 0, 1, 2, \cdots$ and k is an integer, the following relation always holds:

$$\frac{1}{\sin 2x} + \frac{1}{\sin 4x} + \cdots + \frac{1}{\sin 2^n x} = \frac{1}{\tan x} - \frac{1}{\tan 2^n x}$$

(Ref 1966 IMO)

This problem can be solved using mathematical induction. Here we present a complex number based solution.

Let $z = \cos x + i \sin x$. Then it is easy to verify that

$$\tan x = -i \cdot \frac{z^2 - 1}{z^2 + 1} \implies \frac{1}{\tan x} = i \cdot \frac{z^2 + 1}{z^2 - 1}$$

It follows the claim is equivalent to

$$\frac{2 \cdot z^2}{z^4 - 1} + \frac{2 \cdot z^4}{z^8 - 1} + \cdots + \frac{2 \cdot z^{2^n}}{z^{2^{n+1}} - 1} = \frac{z^2 + 1}{z^2 - 1} - \frac{z^{2^{n+1}} - 1}{z^{2^{n+1}} - 1}$$

or

$$\frac{2 \cdot z^2}{z^4 - 1} + \frac{2 \cdot z^4}{z^8 - 1} + \cdots + \frac{2 \cdot z^{2^n}}{z^{2^{n+1}} - 1} + \frac{z^{2^{n+1}} - 1}{z^{2^{n+1}} - 1} = \frac{z^2 + 1}{z^2 - 1}$$

This above relation indeed holds if we evaluate it backwards by noting

$$\frac{2 \cdot z^{2^n}}{z^{2^{n+1}} - 1} + \frac{z^{2^{n+1}} - 1}{z^{2^{n+1}} - 1}$$

$$= \frac{z^{2^{n+1}} - 1 + 2 \cdot z^{2^n}}{z^{2^{n+1}} - 1}$$

$$= \frac{\left(z^{2^n} + 1\right)^2}{\left(z^{2^n} + 1\right)\left(z^{2^n} - 1\right)}$$

$$= \frac{z^{2^n} - 1}{z^{2^n} - 1}$$

Therefore the left side telescopes which will results in $\frac{z^2 + 1}{z - 1}$. This equals the right side of the above relation.

---

10) Let S_n be the minimal value of $\displaystyle\sum_{k=1}^{n} \sqrt{a_k^2 + b_k^2}$ where $\{a_k\}$ is an arithmetic sequence whose first term is 4 and common difference

is 8. $b_1, b_2, \cdots, b_n$ are positive real numbers satisfying $\sum_{k=1}^{n} b_k = 17$. If there exist a positive integer n such that S_n is also an integer, find n.

Complex numbers can also be used to solve other types of problems. For this problem, the following property can be used:

$$|z_1 + z_2 + \cdots + z_n| \le |z_1| + |z_2| + | + \cdots + |z_n|$$

By definition, we have $a_k = 4 + 8(k-1) = 8k - 4$.

$$\therefore \quad \sum_{k=1}^{n} \sqrt{a_k^2 + b_k^2} = \sum_{k=1}^{n} \sqrt{(8k-4)^2 + b_k^2}$$

Let $z_k = (8k - 4) + ib_k$, then

$$\sum_{k=1}^{n} \sqrt{(8k-4)^2 + b_k^2}$$
$$= |4 + ib_1| + |12 + ib_2| + \cdots + |(8n-4) + ib_n|$$
$$\ge |(4 + 8 + \cdots + (8n-4)) + i(b_1 + b_2 + \cdots + b_n)|$$
$$= |4n^2 + 17i|$$
$$= \sqrt{16n^4 + 17^2}$$
$$= K$$

where K is an integer.

It follows

$$16n^4 + 17^2 = K^2 \implies (K - 4n^2)(K + 4n^2) = 17^2$$

This is a standard indeterminate equation. Because $K + 4n^2 > K - 4n^2$, it must hold that

$$\begin{cases} K + 4n^2 = 289 \\ K - 4n^2 = 1 \end{cases} \implies n = \boxed{6}$$

www.ingramcontent.com/pod-product-compliance
Lightning Source LLC
Chambersburg PA
CBHW070249190526
45169CB00001B/345